Shared Obliviousness
in Family Systems

Shared Obliviousness
in Family Systems

Paul C. Rosenblatt

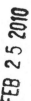

SUNY
PRESS

Published by
STATE UNIVERSITY OF NEW YORK PRESS
ALBANY

© 2009 State University of New York

All rights reserved

Printed in the United States of America

For information, contact
State University of New York Press, Albany, NY
www.sunypress.edu

Production by Ryan Morris
Marketing by Anne M. Valentine

Library of Congress Cataloging-in-Publication Data

Rosenblatt, Paul C.
  Shared obliviousness in family systems / Paul C. Rosenblatt.
    p. cm.
  Includes bibliographical references and index.
  ISBN 978-1-4384-2731-7 (hardcover : alk. paper)
  ISBN 978-1-4384-2732-4 (pbk. : alk. paper)
  1. Family. 2. System theory. 3. Family psychotherapy. I. Title.
  HQ734.R7573 2009
  155.9'24—dc22   2008047582

10 9 8 7 6 5 4 3 2 1

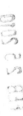

Contents

Preface

Of the many different forms of book preface, this is the type that gives an overview of what is in the book. This is a book about shared family obliviousness, which I see as an important, normal, ordinarily desirable, and sometimes problematic part of family life. The book offers a speculative theoretical analysis of the value of shared family obliviousness, where it comes from, what it does for families, what kinds of problems it sometimes makes for families, and how families may act to deal with the problems that shared obliviousness might create for them.

Chapter 1. Shared Obliviousness as a Family Systems Phenomenon

Shared family obliviousness (the family-wide sharing of lack of awareness and lack of knowledge) is necessary for system functioning. Vastly more information is available than a family system can handle, and information overload makes trouble for families. But why are families oblivious to what they are oblivious to? What kinds of trouble can shared family obliviousness make for a family? How, if at all, can a family overcome shared obliviousness if that seems to be in its interests? The book addresses these questions, and this chapter lays the groundwork by defining terms and establishing theoretical foundations. Shared family obliviousness is different from individual obliviousness because the dynamics of it involve family processes. And obliviousness is different from such related concepts as denial and repression. "Family" is not a concrete reality, but rather a set of relationships that might be defined quite differently from different perspectives, but the explorations in this book of shared family obliviousness are applicable with almost any definition of family. The theoretical starting places for this speculative theoretical book include general systems theory ideas of system control, family systems theory, symbolic interaction and social constructionist ideas about the social creation and maintenance of realities, ideas about the role of shared obliviousness in intersystem relationships, ideas about the links of family systems processes to individual attention processes, and attentional economics.

Chapter 2. Family System Mechanisms for Maintaining Shared Obliviousness

Family systems have many mechanisms for maintaining shared family obliviousness, including family priority setting, establishing standards about what to pay attention to and what not, keeping busy, television watching, avoiding obliviousness-threatening others, rules of etiquette that relate to attention, and various processes of distraction. The rules and patterning of family communication help to maintain shared family obliviousness, and included in that, families may have secrets, selective or distorted notions of their family history, or family myths that help to maintain family obliviousness. In couples, shared obliviousness is promoted by the inclination to partner with someone who is similar to one and by implicit marriage contracts, both of which processes may shape and limit a couple's exploration of their areas of shared obliviousness. Language is also a powerful force for maintaining shared family obliviousness—for example, the terms that are and are not available, the way language can obfuscate or reshape realities. Religion can help to shape and maintain shared family obliviousness, through the ways it shapes attention and through religious processes, including prayer, that can shape and limit attention to various matters. Those with more power in a family have more control over what the family is collectively oblivious to, but perhaps everyone in a family has some power to shape shared obliviousness.

Chapter 3. Family Obliviousness to Context

Families maintain their shared obliviousness in part by being oblivious to many aspects of the context in which they function, including world and national events. Families are pushed toward shared obliviousness by the mainstream media and the U.S. government, educational institutions, and corporations, all of which have their own interests in pushing people toward obliviousness about various matters. Furthermore, families often are oblivious to how their attention, knowledge, information, obliviousness, and ideas of who they are and how to be are shaped by the mainstream media, government, educational institutions, and corporations. But then families may have their own interests in being oblivious to context. For example, they may find life much more comfortable if they are oblivious to the trauma of others, to impending dangers, to historical events that conflict with their realities or raise difficult questions, to the experience of other families, and, for white families, to their unearned racial privilege. This is so even though from some perspectives that obliviousness is not in the best interests of families.

Chapter 4. Obliviousness to Matters within the Family

Family members not only share obliviousness to matters outside the family, they also share obliviousness to matters within the family. Shared (and sometimes unshared) family obliviousness is discussed in this chapter with regard to patterns of conflict, with regard to teen activities and sexual orientation, and with regard to alcoholism or drug addiction within the family. Other topics of family obliviousness explored in this chapter include incest and sexual abuse, obliviousness to the crimes of family members, denial of death and of loss, obliviousness to intergenerational debts, and obliviousness to what is good in the family. The chapter concludes with a discussion of mutual obliviousness in couples and families, the ways that people who are ostensibly closest to each other are also mutually obliviousness to much about one another.

Chapter 5. Shared Obliviousness and Family Decisions

Shared obliviousness may facilitate family decision making. With less information to process, decision making can go faster, and particularly if family members know less about the negative aspects of the decision choice they are moving toward making and less about the positive aspects of the decision choices they are moving toward rejecting, family decision making may go quickly. Furthermore, family information seeking and processing regarding decisions may ordinarily occur in a way that protects areas of shared family obliviousness. And the information environment available to support major decisions may often keep families oblivious to much, because that is in the best interests of those who have the most power in creating the decision environment and who benefit from family decisions being made. Family members may also share obliviousness about aspects of their family decision process, and that too may facilitate family decision making. And after making a decision, family members may work together to be oblivious in ways that enable them to feel good about their decision and that do not let the consequences of the decision undermine areas of shared family obliviousness.

Chapter 6. Family System Responses to Threats to Obliviousness

Families exist in an environment filled with threats to shared obliviousness. How do they deal with these threats? They have ways to minimize threats to shared obliviousness, beginning with limiting family member access to various information sources. Also, just as the larger society has its ways to silence and marginalize dissenters and whistle blowers, a family may silence and marginalize family members who threaten family obliviousness through dissent or whistle blowing. Similarly, if a family member bears witness and attempts to bring awareness to others in the family in a way that threatens shared family obliviousness, the family may well have ways to minimize that person's threat to shared obliviousness. Furthermore, families have ways of screening outsiders who might enter the family domain in order to exclude those who would threaten family obliviousness, and when outsiders do enter the family domain, there are ways to define them, interact with them, and limit them so that they are relatively unlikely to say or do anything to threaten family obliviousness. Sometimes, however, reality intrudes and a family may no longer maintain its shared obliviousness about a matter. But its obliviousness may have worked effectively for years to keep the family from certain kinds of discomfort, and even though a wall of obliviousness is breached, many other walls will remain.

Chapter 7. Obliviousness and Family Therapy

Shared family obliviousness is probably almost never the presenting problem of a family seeking therapy, but family therapists working with a number of different therapeutic frameworks may pay attention to and gear their therapy to matters that are outside of family member awareness. Families in therapy may guard their shared obliviousness in the face of therapy that threatens it, but some family therapists will have as a goal with some client families to help family members deal with key aspects of what they seem to be oblivious to. But then, that obliviousness may be well defended, so the therapist may have to deal with or circumvent family resistance to therapeutic efforts to breach the wall of resistance. Family therapy practice may have its own areas of obliviousness, perhaps including obliviousness to the impact of certain aspects of the larger society on families. For family therapy to be most effective it may be crucial that therapy for certain families take into account the larger societal context or other areas to which therapy practice might ordinarily be oblivious. For families who overcome some kind of shared obliviousness as a result of therapy it may be most likely that the obliviousness they overcome does not open them to so much new information that they risk overload, and sometimes overcoming obliviousness in some area frees a family to be able to handle more information because they are no longer putting so much information processing into struggling with the problem that was linked to the obliviousness they overcame.

Chapter 8. Researching Shared Family Obliviousness

Because almost all research on awareness and its lack has focused on individuals, even in studies of families, researching shared family obliviousness would be innovative. There are, however, a number of challenges to researching shared family obliviousness, including ethical problems, problems in establishing standards for what counts as obliviousness and as shared obliviousness, the risks of creating awareness as an artefact of the research, and the problem of finding a valid nonoblivious standpoint from which to view an area of obliviousness. Furthermore, in evaluating a person's characterization of his or her family as having been oblivious about some matter, there are complexities in evaluating the validity of that claim. Among many interesting possibilities in researching shared family obliviousness, one could research family accounts of their experiences of overcoming obliviousness and one could research the structure of shared family obliviousness (for example, the ways in which some areas of obliviousness are foundational to others). Then there is the obliviousness of researchers. Although obliviousness has its uses in research, perhaps to be a good researcher of shared family obliviousness one must be able to overcome many areas of obliviousness one brings from research practices, the society, and even one's own families.

Chapter 9. Shared Obliviousness That Is
Not Quite Shared or Oblivious

Sometimes what seems to be shared family obliviousness is not quite that. Some family members will have information that others do not have, and that can mean that a family may recurrently struggle to achieve or restore shared family obliviousness. Sometimes the underlying force for family obliviousness may involve a certain amount of awareness by at least one person in the family that obliviousness can protect individual or family comfort, power, privilege, or image. Sometimes family members seem to collude in trying to maintain obliviousness about some manner, which implies that they know something even when they are trying not to know. Sometimes families occupy a place between obliviousness and awareness in which family members sample or skim information ("the headlines") on certain matters, but do not usually go beyond succinct overview information. Sometimes families have ways of monitoring information to which they are ostensibly oblivious, so they may become aware, even while seeming to be unaware, of certain kinds of threats, opportunities, environmental changes, changes in the family, and so on. Perhaps many families are organized so that someone in the family is an information specialist in a key area of ostensible family obliviousness. The family may downplay or seemingly ignore the knowledge the information specialist has, but it may also make use of that information if that becomes appropriate. Another way in which a family may seem to be oblivious but may not be is that sometimes family members know more than they can say, and that knowledge influences the family even though it cannot be verbalized. In addition, families may keep records that include information to which they are currently oblivious, but the records provide the potential to overcome obliviousness.

Chapter 10. The Future of Shared Family Obliviousness

Families may be oblivious to potentially useful hints, predictions, and projections about the future, and so they may be less well prepared than they might be for the future. In the future, they may be judged by their children and others who depended on them as having failed to deal adequately with a future they should have anticipated. It is possible, however, that some families can process more information than they are now doing, and so they may be able to become better prepared to deal with the future. On the other hand, they may be pushed toward being less prepared than they could be by what is and is not available in the information environment. What is in the future is of course unknowable, but it may include radical changes for families—for example, in what constitutes families or in how many people share a dwelling. The future changes may push families toward greater shared obliviousness, but then families may do better who allow some family diversity in who is aware of what. In a sense the future of shared family obliviousness is about all of knowledge and all that goes on in families. So attending to shared family obliviousness gives us a window to everything. And that brings us to issues of teaching about shared family obliviousness. Teaching in the social sciences and psychology may often have the potential to penetrate into areas of shared family obliviousness. Perhaps any possible penetration into areas of family obliviousness has limited effects, because family obliviousness is well defended. But it is also important to acknowledge that teachers are not employed to make trouble for students and their families. So teaching in areas that could impinge on the shared obliviousness of a student's family should not be about changing the student and the student's family, only about giving the student conceptual tools. What the student does with the tools is up to the student, not the teacher.

This work of theory about shared family obliviousness is a beginning. There is room for considerable theoretical development and research in this area. As one step toward that, this chapter offers a list of various patterns of family obliviousness discussed in this book. Future research and theory may explore these patterns and the processes of moving from one to another, of maintaining shared obliviousness and of moving into shared obliviousness.

Finally, although not a self-help book, the book concludes with a discussion of key areas to consider if one is seeking to make sense of, come to terms with, or change a pattern of shared obliviousness in one's own family—simplifying, specializing, and being aware of the family system roots of certain forms of potentially harmful shared family obliviousness while accepting that shared obliviousness is often desirable and necessary.

Acknowledgments

My work on this book has benefited from stimulating comments, questions, suggestions, and observations from many people. I thank you all, and I want to give special thanks to Sara E. Wright, Jim Maddock, Liz Wieling, John Barner, Rob Palkovitz, Hilary A. Rose, Bill Allen, Marshall Lev Dermer, Clinton Gudmundson, Julie Zaloudek, and Julia Malia. None of these people should be blamed for the shortcomings of this book. They all did their best to put me on the right path.

Shared Obliviousness as a Family Systems Phenomenon

IMAGINE A FAMILY IN which nobody is aware that their home is in grave danger from a flood that is bound to happen in a few days, that the value of their home and all the homes in their community is plummeting because the local and national real estate markets have been built partly on chicanery and illusion, that their religious leader is a sexual predator, or that the family history that is important to them and that they believe to be true is mythical. Imagine a family in which nobody realizes that Dad is an alcoholic, that their recurrent arguments are patterned in ways that prevent them from resolving any disagreement, that the grandmother who they think is perfectly healthy had a stroke last week, and that their favorite political leader systematically lies. Imagine a family in which nobody knows that he or she has been silenced in many ways by the way the family interacts. These forms of shared obliviousness and others like them occur in all families, because all families share obliviousness to a great deal. By obliviousness I mean a state of being unmindful or unaware of something, of being ignorant or not conscious of its existence. By shared family obliviousness I mean the family-wide sharing of lack of awareness and lack of knowledge. This book is not about what family members know but do not talk about. It is not about what they know but wish they did not. It is not about what they know but are too embarrassed to claim to know. It is not about what they know but are not sure about. It is about shared areas of inattention and ignorance.

The members of any family share obliviousness to vast amounts of information, and they must do so. But that does not mean obliviousness about any specific matter is good for the family. Sometimes a family's obliviousness can make serious trouble for it. Because families are oblivious to so much and because their shared obliviousness can make serious trouble, I believe that we cannot understand families, how they function, how they get into trouble, how they solve or do not solve problems, or how to help them without understanding the ubiquity and dynamics of shared family obliviousness.

This book is intended to open a new line of theorizing about families and family systems. It builds on many theoretical traditions, but most of all on family systems theory, and offers extensions to that theory. The topic of shared family obliviousness is new ground for the family field and the social sciences. There are works about individual awareness and unawareness and works about family members sharing values, a culture, and the like, but no published work that I know of has pulled that material together, provided a framework that covers the disparate areas, and opened up the broad area of shared family obliviousness for theoretical analysis.

The Necessity of Shared Obliviousness in Family Systems

Family systems, like other information processing systems, cannot possibly function or function well if the system as a whole and its members as individuals try to keep track of too much. There is an infinity of information available. Even a minute amount of all the information that is potentially available would be far too much for a family to process. And most of the available information is useless and would be distracting in terms of what a family and its members need or want to know or are capable of knowing. So the system must set priorities about what to attend to and what to ignore. In the sense that obliviousness to a vast amount of information means that a family has succeeded in prioritizing information so as not to be overloaded, shared obliviousness is an achievement of family system functioning.

How does a family achieve shared obliviousness? How does it come about that this is what the system is oblivious to and that is what it attends to? How is shared obliviousness built up and maintained? Presumably, family obliviousness is nurtured, demanded, and regulated by family system processes, including explicit and implicit family rules. That is, families are not randomly oblivious. Family system mechanisms are in operation that define what it is that family members should detect, attend to, know, perceive, talk about, remember, and respond to and what they should ignore and be unaware of. And these mechanisms are linked to societal rules and forces—including societal values, social pressures, what is in the mass media and in education systems, what is talked about in social circles in which family members travel, and what is important in family religious traditions and religious services.

However necessary shared obliviousness is, it can have its costs to a family. Shared obliviousness may block adaptations, interfere with connections with other systems, and keep families ignorant in ways that prevent them from reaching goals that are important to them or that are even necessary for the continued existence of the family and individual family members.

Defining Shared Obliviousness

Obliviousness can be defined as a state of being unmindful or unaware of something, of being ignorant of it or not conscious of its existence. When obliviousness is shared in a family, the family members will individually and collectively distance, avoid, lack interest in, be unaware of, or lack engagement with relevant information, perspectives, meanings, interactions, places, memories, and events. Almost always, shared family obliviousness does not represent a choice by anyone in the family or by the family collectively. Shared family obliviousness just happens. And in that obliviousness they are unconcerned about (in large part, or quite possibly entirely, because unaware of) whatever it is they are oblivious about.

Shared obliviousness is a property of social systems. All social systems tune out a great deal of available information as they function and work toward what seem to be their goals. Systems can do this through a summation of individual obliviousness and through organizing in such a way that the system and everyone who is part of it is oblivious. This book is an extended commentary on how one category of systems, families, can organize to be oblivious. That organization includes values about what counts as interesting and important. It includes education that focuses family members here and not there, walls (literal and metaphoric) that block off certain information sources, and system-wide rules that define only certain information sources as worthy of attention.

General systems theories typically do not problematize inputs but assume that inputs are so obvious and can so be taken for granted that there is no reason for a system analyst to explore why it is that of all the potentially accessible inputs a system only detects and makes use of the ones it does. Similarly, in the information systems literature, information might be defined as any stimulus that has changed recipient knowledge (e.g., Lawrence, 1999, p. 2). By contrast, the concept of shared obliviousness introduces the notion that systems at some level must always select and filter information. They must always have processes for separating what to attend to from what not to attend to. Understanding the bases for those processes would tell us a lot more than simply assuming that inputs are whatever they are or that they exist if something changes in the system in response to them. Inputs to systems should not be taken for granted. It is better instead to raise questions about how it is that the system takes in or reacts to this and not that.

Shared family obliviousness does not necessarily involve a lack of focus or absorption. Indeed, an important process of achieving obliviousness about some matters is to be focused on and absorbed in other matters. Hence, an important reason for a family system to focus on this or that is that it is then much easier for it to be oblivious to other things. For more on this, see chapter 2.

Shared family obliviousness often happens passively. Family members live their lives and only encounter certain information. They may not resist information that comes to them, but often, as far as they are aware, it is only certain information that comes to them. But then how is it that some potential information is welcomed and much is treated as though it is nonexistent? How is it that some things count as information and some things do not? At any one instant there are trillions of bits of information immediately available or easily reached. People hear, see, smell, feel, taste this or that, but tune out vastly more than they tune into. Something has sensitized them to count some things as signal or information and other things as noise or noninformation.

Shared Obliviousness versus Individual Obliviousness

Although in the dictionary definition, in ordinary English language usage, and in everyday conversation and writing, obliviousness is seen as an individual phenomenon, this book focuses on shared obliviousness, on obliviousness as a family property, occurrence, experience, and achievement. Individuals are always in social systems, including family systems, and they are rarely, as this book argues, oblivious on their own. They are almost always oblivious with others in a system or oblivious in part because others in a system in which they participate are oblivious or want them to be oblivious. Thus, any analysis and critique of an individual's obliviousness that focuses only on the individual misses how much the obliviousness is linked to what goes on in the individual's family and in the other parts of the individual's social environment.

If we only look at individual obliviousness, we will not understand how much individual obliviousness is created by, maintained by, and in service to the family system. That is, often it is in the family system that the matters about which individuals are oblivious are targeted for obliviousness, and it is in the family system that much goes on that maintains individual obliviousness. And that obliviousness often has a value to the family system in maintaining the arrangements, rules, power structure, comforts, patterns, images, etc. that it has as a collectivity and that the individuals in the family share. Additionally, obliviousness in a family system may at times be linked to the deepest, most difficult, most daunting, most disruptive, most divisive, most hard to resolve, most painful relationship problems in that system. Even obviousness to something outside of the family system may be about matters inside the system. For example, if members of a white family share obliviousness to racism outside the system, that could be in large measure about dynamics within the family—perhaps shared aspirations for status or shared discomfort with the idea that the family has gained immensely from unearned privilege because of being white.

Shared Obliviousness versus Denial

"Denial" overlaps with obliviousness in that with denial, as with obliviousness, people have tuned out information. But with denial there is an ongoing rejection of information that is available to people if they only were to attend to it, and this is information that at some time and in some way they knew existed. With shared family obliviousness there may not ever have been an awareness of the information to which the system is oblivious. Following Cohen (2001, pp. 7–9), people may deny that something is true; they may deny certain interpretations of what they accept as true, or they may deny the moral, psychological, or political implications of what they accept as true. For example, a family that denies that a family member is an alcoholic will have the bouts of drunkenness, the alcoholic failures to follow through on commitments, the empty bottles, the smell of alcohol, and so on in front of them recurrently. One could say that they are denying the evidence that these things are happening. One could say that they are not denying evidence but are denying the interpretation that the evidence means that alcoholism is present. Or one could say that they accept that the evidence exists and even label the evidence as clearly indicating that "alcoholism" is present, but they deny the moral implications, the psychological implications, or the implications for the family and its internal workings. In any case, by using the term *denial* we emphasize how they must rather actively persuade themselves that things (facts, interpretations, implications) are not what they would seem to be to an observer who is open and not defended against the available information.

I differentiate shared obliviousness from denial in that I think often families are oblivious without there ever having been the active elements implied in denial. There was, for example, never awareness in the family of anything concerning a family member's alcoholism or even of the concept of "alcoholism." Or if there was awareness of these things, there was never awareness of the meaning of what the family member did as representing alcoholism.

However, I can imagine situations in which shared obliviousness may arise from active denial mechanisms. That is, shared obliviousness can involve active efforts not to know, be aware, perceive, understand, interpret, remember, etc. Then it is not that family members only process the information that is available to them but that they actively work at keeping away from certain information or at not processing certain information. Perhaps it is too uncomfortable for the family to think that Dad is an irresponsible drunk, so they do not (either because they avoid certain information about him or because they avoid interpreting that information). That raises questions about seemingly passive shared obliviousness. Is passive shared obliviousness sometimes a product of family members deciding not to seek or make sense of certain information? Is passive obliviousness at times a point in a process that at one time involved active choices about what to ignore or put aside? And is each family member on her or his own in being passively ignorant, or might others in the family help them toward that passive ignorance? Perhaps there are times when a family member becomes oblivious in order to please others in the family, to avoid family criticism or hostility, to honor family standards, or to be a proper kind of spouse, parent, sibling, offspring, grandchild, etc. The active versus passive distinction provides a point of differentiation between denial and obliviousness, but one could say that some areas of shared obliviousness involve denial or are based on past denials.

Shared Obliviousness versus Repression

"Repression" (Freud, 1924) can be understood as the psychological act by an individual of excluding memories, desires, feelings, wishes, or fantasies. From my perspective, many areas of shared family obliviousness have no relevance to Freud's notion of repression. These are areas that never have been matters of awareness, interest, emotional meaning, or psychological importance. For example, a family may share obliviousness to what for them is psychologically unimportant about politics in a distant country, the history of geology, or the meals eaten by their neighbors.

Freud certainly had a sense of larger social processes, for example, in *Group Psychology and the Analysis of the Ego* (1960), but the Freudian concept of repression rests on mechanisms of individual psychology. As such, one could make the case that Freudian repression is an engine that drives individual obliviousness in a system where the areas of shared obliviousness are ones in which people have tuned out information that once was known to some extent, that has great and challenging psychological meaning, and is therefore potentially of substantial importance. With "repression" and the forms of individual obliviousness that might be driven at least in part by repression, the information that is tuned out is never far from being available to know. With repression the mechanism of tuning out, at least in classic psychoanalytic thinking, involves active work that potentially could leak information (through errors and dreams, for example) about what was repressed. That could be true for some areas of shared family obliviousness (for example, when family members seemingly share obliviousness to sexual abuse within the family). But with most areas of shared family obliviousness, there never was awareness or meaningfulness. So what family members are collectively oblivious to typically never had a meaningful place in memory, in thought categories, or even in language to put the information in if any family member should ever move toward giving up the obliviousness (cf. Bowker, 2005; DeVree, 1994). With repression and perhaps with areas of shared family obliviousness that arguably rest in part on individual repression, should the material ever become unrepressed, there would ordinarily be extremely meaningful, however uncomfortable, places to put the information. But I think most of what moves out of shared family obliviousness and into shared family awareness does not have much meaning to the family. So sometimes shared family obliviousness may be quite a bit like repression, but I think typically it is not.

Who Is to Say What Obliviousness Is?

Given that obliviousness can only be identified and reacted to if someone is not oblivious, the questions arise, "Who is to say what obliviousness is?" "Who is to say when there is shared family obliviousness?" From a postmodern perspective (e.g., Rosenau, 1992), there is not a solidly objective position from which to evaluate the world. One person might claim that obliviousness is present, while another sees things differently. Person A says something was overlooked; person B says there was nothing there to overlook. Person A says something was forgotten; person B says there was never anything to forget, or embedded in what is remembered is all that A says is forgotten. Person A says that a family is oblivious to something very important; person B says that A is concerned about trivia. Whose perspective counts? Who is correct? What are our criteria for deciding what is correct? Are these questions resolvable? If we believe they are, then a case can be made that a key issue for systems thinking is the question of who has standing to judge which viewpoint is correct or relevant (Flood, 1999, p. 70).

Related to the question of who is to say what obliviousness is, to say obliviousness is present one must be good at detecting what people are aware of, know, perceive, and think about. And this requires some location for observation, but any location for observation may be embedded in a system that creates, shapes, and sustains its own obliviousness. Moreover, some locations come with vested interests such that judging others to be ignorant or oblivious serves the economic or other interests of those doing the judging (Hobart, 1993). So in the perspective of this book, an observer's location is always open to question as a source of valid information untainted by the observer's own obliviousness. The detectors of obliviousness are inevitably limited by their own obliviousness and the systemic embeddedness, biases, values, and goals that are served by that obliviousness.

Obliviousness and Awareness Are Linked

Obliviousness is produced in systemic connection to what is attended to. In a sense, each needs the other. We could not be oblivious to a, b, and c if we were not looking at x, y, and z. We could not look at x, y, and z if we were not oblivious about a, b, and c.

From another perspective, obliviousness is always about difference, about what information is to be perceived or thought about or known versus what is not. For this difference to exist, there must be some awareness of the difference (Bateson, 1980, p. 76). Once something counts as information, other things do not. People might not have any reason to know much about what does not count as information, but at the very least by knowing what does count they have some sense of what does not, even though they might not know specifics about any of the things that do not count.

From still another perspective, a case can be made that obliviousness and awareness are at times in dialectical relationship. Discomfort with obliviousness or the realization that a certain area of obliviousness is not good for the family can motivate family discovery, curiosity, and learning. More generally, awareness of ignorance can lead to new knowledge (Schneider, 2006). The dialectic can also work the other way, with awareness and knowledge somehow pulling a family toward obliviousness—for example, if some sort of knowledge is too uncomfortable to deal with. The continuing discomfort may be in ongoing relationship with continuing efforts to be oblivious. And to the extent that the concept of dialectic calls for a synthesis of obliviousness and awareness we might look at situations where people work for superficial knowledge, for example, the skimming of headlines referred to in chapter 9. That is, resolution of the tension between knowledge and obliviousness might involve a synthesis that is part way between complete obliviousness and full knowledge, perhaps superficial knowledge or a commitment to knowing only the major, overarching information.

Shared Family Obliviousness as a Metaphor

One can take my usage of shared obliviousness as applied to families as a metaphor that draws on what we know about the obliviousness of individuals. Ordinarily, when English speakers use the term *oblivious*, it is the individual who tunes things out, is totally ignorant, even to the point of not being aware of his or her ignorance. As a metaphor, this usage of the term *oblivious* or *obliviousness* in writing about shared obliviousness highlights some matters and obscures others (Lakoff & Johnson, 1980). It highlights that families are like individuals in clearly being aware of some matters and clearly not being aware of others. It draws on the entailments of individual obliviousness to suggest that certain things may go on with shared family obliviousness, for example, selective perception, having no interest in learning certain matters, turning a deaf ear to this or that, not giving a thought to certain matters, and lacking curiosity about those matters. However, to use the term *oblivious* or *obliviousness* metaphorically obscures that a family consists of diverse people, whose attentional interests, capacities for curiosity, remembering, and nondefensive information processing are likely to be diverse. From that perspective, all the members of a family may share obliviousness to some matter, but that obliviousness may be accomplished through family dynamics that swamp the individual diversity in the family. The metaphoric application to families of a concept usually applied to individuals also obscures how much family members may be diverse in what brings them as individuals to shared obliviousness to some matter, and they can also be diverse in how much they push on other family matters to be oblivious to these matters. It can be one person's anxiety, defensiveness, rage, hunger to learn, curiosity, moral outrage, or accidental learning that can drive what a family shares obliviousness to and what it is not oblivious to.

Defining Family and Family System

Defining family is not simple, since many people, both the general public and scholars who focus on families, seem to be oblivious to how people define family as they live their everyday lives (Gubrium & Holstein, 1990). Defining family is also subject to hot political debate these days—for example, as it applies to same-sex couples or to whom an employee can designate as family for purposes of family medical benefits. Complicating efforts to define family, there is an enormous diversity of families within and across cultures and also an enormous diversity of conceptions of family across cultures and history. I believe the analyses in this book hold up no matter how family is defined, as long as the definition excludes groups of people who have no interactions. I do not assume that couples in families must be heterosexual or married to count as part or all of a family. I do not assume that a family must share a house or must be multigenerational. I do not assume that members of the family see the family in the same way or that family members agree on who is in the family or whether they are a family. I do not assume that families are egalitarian or patriarchal or dominated by one person. And even though I write about shared family obliviousness, I do not assume that family members see things in the same way or know the same things. In fact, it is a central characteristic of families that family members are diverse in opinions, knowledge, awareness, memories, and much else. So this book focuses on shared obliviousness but also pays a great deal of attention to situations in which family members do not share obliviousness or come to shared obliviousness from the same place.

A family is a system because of the interdependency and connection of at least some of the members and because there is a certain amount of order and pattern to what goes on among its members (see Rosenblatt, 1994, p. 51). The patterning, though recurrent, also may change (because situations change, individuals change, relationships change). In focusing on the system, the focus is on interaction and process, not separate individuals and not a personified family that thinks, feels, acts, and remembers as though it were an individual. The focus on family system is then on what goes on between and among family members. So this book on obliviousness in family systems focuses a great deal on how shared obliviousness is shaped, determined, and often shared as a result of the interconnections of family members. It also focuses on how the differentiation in awareness and knowledge is shaped and determined as a result of the interconnection of family members.

In saying that a family is a system, I am not saying that family members are in agreement about who is in the family. Family members may often differ about who is in the family or whether they are a family. The outside analyst offers one view of the system, not the only view. And if an outside analyst relies on the views of one or several family members in order to build up a view of a family, that does not mean that is the only view that could be derived from how family members characterize the family. So when it comes to thinking about obliviousness in a family, perhaps the members of the family share many areas of obliviousness, and perhaps they are in agreement on a number of matters, but there will always be areas of difference. Family members may be oblivious to some of their differences, assuming that they are similar or alike when they are not. But they will also have differences that they are aware of, and some differences will be matters with which they struggle (Gilbert, 1996). Family members may also be oblivious to their interconnections and to the ordering and patterning of their relationships, which is interesting from the point of view of trying to understand the family but also challenging from the point of view of using what family members say in order to understand the family.

First Theoretical Tools for Thinking about Shared Family Obliviousness

System Control

From the perspective of family systems theory (e.g., Rosenblatt, 1994), systems need order. Without order there is no system and quite possibly no capacity to maintain an adaptive fit to physical, social, symbolic, etc. environments (Buckley, 1967, pp. 164–206; Kantor & Lehr, 1975). Family systems operate with mechanisms that maintain order (Kantor & Lehr, 1975; Rosenblatt,

1994, pp. 128–51). Among those mechanisms are explicit and implicit rules about what family members may or may not do and feedback loops that stop deviations from accepted patterns (quite possibly including chiding, correcting, threatening, or even expelling rule-violating family members). Shared family obliviousness is an important mechanism for maintaining order. It can be, for example, a defense against information that would disrupt family order. It can be a shield against what is outside and a protection of what is inside the family. It makes it difficult for certain kinds of outside information to enter the system and then rock the boat or change the system. This does not mean that system order is necessarily benign or that the information from the outside would be harmful to family members. It only means that systems tend to maintain the order they have, whether that order is good or not by some standard. One might think, for example, that once the members of a Florida family understand that there is a good chance that global warming will lead to the destruction of coastal and southern Florida in their lifetime, they could hardly maintain their current system of mundane priorities, plans, investments, and activities. So a control system that pushes for shared obliviousness to global warming maintains the family system as it is and blocks it from taking global warming as a serious threat.

One framing that some general systems theories offer is the idea of systems as about information (e.g., Bertalanffy, 1968, pp. 41–44; Frick, 1959), the transmission of information, the exchange of information, limits on where information goes, and inputs of information being transformed by the system into informational outputs. If we think of family systems in terms of information, we can also think that family information systems only work well if they are not overloaded (Paolucci, Hall, & Axinn, 1977, pp. 116–17), and if they manage information processing effectively. Thus, information processing would ordinarily require ignoring substantial amounts of information. As Frick (1959, p. 614) wrote, in information theory "information, and ignorance, choice, prediction, and uncertainty are all intimately related." As he saw it, information processing is limited by the capacity of the system to make use of information. From that perspective, if a family system is not attuned to information about how the choices being made about where to live are bad for the health of people in the system, it cannot make use of the information, however plentiful and accessible such information seems to others. Or consider, to take another kind of example, family systems as being about genetic information. Genetic information is combined to make new humans. Genetic systems only work because they are closed to all sorts of information that is irrelevant or harmful, so in a sense the biology of DNA, gametes, and reproduction hinges on the obliviousness of reproductive systems to almost all the information that is out there. Genes only use the DNA they are organized to use in the ways they are organized to use it, not all the other DNA and combinations of DNA and not the other combinations of proteins that fill their environment. The information that is kept out would, if allowed in, create disorder or undermine the system's functioning. Extending the analogy to family systems, all family systems maintain system control by being closed to a great deal of information, and they must do this in order to function.

Congruent with the idea that systems must maintain system control through obliviousness, Frick (1959, p. 615) asserted that a system that is open to a greater range of information faces more uncertainty, and that uncertainty requires even more information and information processing. For example, if a couple considering where to settle in retirement knows a great deal on many dimensions of importance to them about a thousand different possible retirement locations throughout the world, their lack of obliviousness will make it difficult to come to a decision. If they only consider two locations and do not know much about them, their decision will be much easier. So obliviousness helps to make easier and quicker decisions. It does not necessarily make wiser decisions or the best possible decisions by all criteria, but it makes things happen.

Another perspective on system control through obliviousness comes from the Corning and Kline (1998) view that information is not a thing in itself but an aspect of the relationship between things. So obliviousness is, in that sense, nonrelationship. Thus, if a system is oblivious to certain events or phenomena there is no capacity in the system to control or use whatever comes from those events or phenomena. And conversely, if a system has no capacity to use or control a certain body of information, then obliviousness to that information makes perfect sense. On a related line of argument, thinking of family systems holistically demands that we understand how obliviousness is meaningfully linked to the other elements of the systems. Klinger (1977, pp. 42–43) argued that a primary mechanism for selective attention by individuals and hence for individual obliviousness is the importance or interest value of what is attended to. From that perspective, people are oblivious to what they think does not matter much or at all to them, and hence if something becomes important for some reason it becomes a new source of selective attention. As Klinger saw it, there is an attentional mechanism which he called "preattentive processes" (Klinger, 1977, p. 44, citing Neisser, 1967). These processes represent motivation, interests, and related matters that make some things important to attend to and others not and apparently are in operation while attention seems fully focused somewhere else or, as in sleep, nowhere at all (Klinger, 1977, p. 46). Extending Klinger's analysis to families, we would expect family obliviousness to be linked to shared family values and goals, and these "preattentive processes" are central to the attentional control system of families. (For more on this, see chapter 9.)

In the Kantor and Lehr (1975) analysis of family systems, there is a kind of obliviousness in the sense that information remains unimportant as long as family members do not recognize it as being relevant to the family (p. 40). Furthermore, in the Kantor and Lehr analysis, the information that counts is information that regulates emotional and physical distance among family members. Since, in their thinking, each family member is also a semiautonomous system, it is possible for family member A to bring information into the life of other family members, because for family member A the information has some interest. Once it is brought into family interaction, it can become important if it is shaped to fit into and become part of the distance regulation of the system.

A final way to introduce the idea that obliviousness and system control are linked is to consider the concept in systems thought of structural coupling (Maturana & Varela, 1987, p. 102). It is the idea that a system and its environment are a unity and must co-evolve. From that perspective, obliviousness in a family system co-evolves with aspects of its environment. So, for example, if shared obliviousness to racial, class, and other privileges exists in a family, it co-evolves with segregation in neighborhood, workplace, religious congregation, local schools, and health care, so that, among other things, the privilege never has to be questioned or challenged. It also would co-evolve with a political and economic system that maintains the other parts of the ecosystem that co-evolve with the shared family obliviousness. So the control aspects of a family system are not only inherent in the family system. They are also inherent in the systems with which the family is coupled.

Shared Obliviousness and the Analysis of Family Systems

Another relevant area of systems thinking is "critical systems heuristics" (Ulrich, 2002, 2003). In Ulrich's (2002, 2003) perspective, "boundary critique" is at the core of critical systems heuristics. It is the idea that decisions are always made about which facts and norms are considered relevant in systems analysis and which are not. What is included within the boundaries is the basis for rationality, systems analysis, and everyday life. What is considered outside the boundaries is generally ignored. This selectivity, Ulrich wrote, merits critical analysis in order to understand what underlies the selectivity and to explore the ethical issues involved. Thus, boundary critique may lead to a challenge of boundary decisions and to competing ideas of what should and should not be brought within the boundaries. By bringing in new facts or values, we can reach new boundary judgments, and these judgments may often lead to new benefits as previously ignored system pieces and phenomena are included and as those people who were not benefited or were even harmed by the prior boundary judgments possibly find their realities and needs counting more with the new boundary judgments. Underlying critical systems thinking is sensitivity to the social and political power issues underlying one systems analysis versus another (Midgley, 1996). I think the analysis this book offers of shared family obliviousness is in the spirit of critical systems heuristics in challenging systems views that do not pay attention to shared obliviousness.

Family Obliviousness and the Construction of Reality

From the viewpoint of social construction theory and symbolic interaction theory, one can think about a couple or family as constantly in the process of constructing and reconstructing reality together (Berger & Kellner, 1964; Berger & Luckmann, 1966; Rosenblatt, 1994, p. 157). In this view, the key to reality construction is talk. By talking together, couples and families talk the day's events at home and at work, the events in the news, the meaning of their meal together, and so on into reality. Part of the reality they create has to do with importance. By not talking about X and Y but talking about Z, they are saying X and Y are not important, but Z is. There is only so much family or couple time for talk (in fact, often very little—Rosenblatt, 2006). So reality construction in a family or couple requires obliviousness to a great deal.

In the process of reality construction, a couple or family selects and organizes information, symbols, and concepts in order to create realities that have some coherence and consistency. That means that they must put aside and ignore information that does not fit those realities, and they will not seek information inconsistent with those realities (Rosenblatt & Wright, 1984). However, some of the information they have put aside because it is inconsistent with their constructed realities will be in the shadows, not lost (Rosenblatt & Wright, 1984). Thus, if their constructed reality is that they are a good and loving family, they may have in the shadows information about the times they have been nasty to each other, bored by each other, emotionally distant, and disgusted with each other. If they remain even somewhat aware of inconsistent or contradictory information, that information threatens to undermine their constructed realities. One would expect that they would work at being oblivious to what does not fit their current constructed realities. However, to the extent that the inconsistent and contradictory information still exists in memory or in their everyday interactions, it may at some time lead to an abrupt flip out of the current reality and into a new one that fits what has been in the shadows (Rosenblatt & Wright, 1984). Then they might have the reality that they hate each other, are emotionally distant, and do not love each other, and what they would then work at being oblivious to and keeping in the shadows is information inconsistent or contradictory to that reality.

Family Obliviousness as a Defense against Other Systems

A family system exists in a world of many systems that are or could be in contact with the family—for example, the systems of employers, schools, and neighboring families. The family and these other systems may at time be in conflict for resources (for example, a family member's time). They may at times be in conflict over competing views of reality (for example, that the good life involves dedication to one's family versus the good life involves dedication to one's work). Certain aspects of shared family obliviousness may be understood as strategic acts in these conflicts. If, for example, a family's obliviousness means that the family members tune out the standards, claims, and arguments coming from another system, that helps the family to maintain its own standards. The obliviousness closes the family to, for example, hostile ideas and bits of information from the other systems that could undermine commitment to the realities of the family's current system. From a family systems theory perspective (Rosenblatt, 1994), one can understand the defense as a matter of establishing and maintaining the external boundaries of the family. If a family cannot resist the forces for change coming from outside, at the very least it cannot remain stable. Perhaps it cannot even continue to exist as a system of meaningful, organized relationships.

Family Obliviousness and What Members Are Set to Sense

Obliviousness in families can be understood as a product of what it is that people are organized in their families to sense. For example, we humans are generally unaware that gravity waves generated in deep space are constantly passing through our bodies (Blair & McNamara, 1997). Our bodies did not evolve to detect gravity waves, and our families have not given us any reason to build and use artificial sensors to detect gravity waves in our everyday life. So we are not organized to sense them. On the other hand, many families have organized to detect the dangers that are meaningful to them in the environments in which they function. For example, if a family has decided that the members are vulnerable to drive-by shootings, any family member might be instantly aware that it is evening and a car with its lights out is moving slowly down the street toward the family house.

From the perspective of Goffman's *Frame Analysis* (1974), people are oblivious to what is not in frame. The frame defines what it going on, what is important, what has meaning, and what language should be used to talk about what is going on. Whatever is out of frame is irrelevant. Frame and the selective attention that is part of it are constructed out of sociological processes, including, to a very important extent, family conversation (Berger & Kellner, 1964; Berger & Luckmann, 1966; Rosenblatt, 1994, p. 157). From a frame analysis perspective, while one is in a frame established by family it would typically be strange and meaningless to leave the family frame behind and to move into a very different frame. However, in Goffman's view, it is certainly possible to move to a new frame, a new definition of the situation, and that could open up areas to which one has been oblivious. It also seems possible for frames to clash in a particular setting—for example, for people with very different views of something to debate with each other or for an individual to be torn between, say, a family frame and an occupational one. But even with frame switching or situations where more than one frame is functioning, I think it is safe to assume that there still is an enormous amount that would remain matters of obliviousness to all the frames in play.

Obliviousness and Attentional Economics

There is vastly more going on in the world and even in a family than any individual or an entire family can attend to. Of necessity, there must be some sort of attentional economics operating to select what is attended to versus what is ignored (Bateson, 1972, p. xv; Schneider, 2006; Thorngate, 1988).

I believe that the chaos that comes with trying to process too much information is something that the members of many families have encountered. The family at times attempts to take on too much information, and then problems arise. Communication errors occur, misperceptions and errors of judgment increase, important information is missed, possessions are lost or misplaced, high priority chores fail to be done, things are forgotten or overlooked that are much too important in the family priority system to be forgotten or overlooked, accidents happen. At this point family members might quite possibly become conscious that they have tried to process too much information and that they must do something to lighten the information load. I think that many families will have had such experiences and will have family values that give importance to something like attentional economics, to keeping things more simple than they might be.

In a rational attentional system, choices would be made in terms of likely gain from paying attention to this versus that, or perhaps there would be a more basic principle of attending in order to minimize harm, damage, or threats to survival. But there is so much to attend to that it is impossible to know even the littlest bit about more than a very small fraction of it all. So I think it is impossible for a family system to be perfectly rational in sifting and evaluating all possible choices of what to attend to.

Also, what can seem to be about attentional economics is often, as I argue more fully at a number of places in this book, not about attentional economics but helps families avoid what is threatening or dangerous, uncomfortable, disruptive of a wide range of routines, or capable of making family members feel guilty, ashamed, or diminished. Connected to this is the question of whether, when, or how often decisions to attend to this or that and to be oblivious to other things are made consciously or represent unconscious processes. I would imagine (following Cohen, 2001, pp. 3–6) that these decisions are sometimes made consciously by at least some members of a family, sometimes made unconsciously, and sometimes a blend (for example, family members may choose consciously not to learn enough to have to make a conscious decision about whether to attend to information about how many children starve to death in the world each day). I would also think that what is conscious may become unconscious and vice versa. I am sure, for example, that there are conscious mechanisms that at times move family members or a whole family from obliviousness to attention. That is, even if there were unconscious processes underlying a family's obliviousness to, say, global warming or their own heterosexual privilege, events could occur that would make paying attention to these matters a fully conscious act.

2

Family System Mechanisms for
Maintaining Shared Obliviousness

Shared family obliviousness does not necessarily just happen. In part it arises out of and is maintained by substantial family effort. On the surface, a family that is oblivious to something may seem simply to be oblivious, with as little conscious effort by family members as it takes for them to blink or breathe. And that is no doubt true for most areas of shared family obliviousness. But for some areas of shared obliviousness there probably was in the past and may still be substantial family work to build and maintain the shared obliviousness.

At least some family members may have a sense that dropping shared obliviousness about a matter may lead to great unpleasantness. For example, they may have a sense that there is something in the history of how Grandfather acquired his wealth that would be unpleasant to know, and so without knowing what it is they are oblivious to, they work together to share obliviousness about the matter. Or some family members may have a sense that understanding the details of what is in their food and how it got there could be horrifying, and so in that topic area, even though they do not know what they are oblivious to, they work hard to maintain shared obliviousness. I believe that in any family there are enough such areas that shared obliviousness is not a casual add-on to a family system. It is often central to the communication, interaction patterns, rule setting, daily routines, holiday celebrations, satisfaction, comfort, recreational choices, and content of conversation that are at the core of family system functioning. There are no doubt an enormous number of areas in which shared obliviousness is just there, without anyone in the family feeling or thinking anything remotely close to the topic area.

But there are also, I think, quite a few areas where there is enough of a hint of threat that there is to a certain extent active effort in the family to maintain shared family obliviousness.

One might suppose that the family system mechanisms for maintaining shared family obliviousness would be deeply buried in the locked vaults of the individual unconscious, in the forgotten past, or in events in the here and now that are so subtle that they are almost impossible to detect. But some of the key mechanisms for maintaining family obliviousness are on the surface in the everyday lives of families and easy to see. That is, what goes on day in and day out in every family is partly about maintaining shared obliviousness. This chapter addresses some subtle and relatively hidden mechanisms for fostering shared obliviousness, but it begins with what is there for anyone to see, priority setting, socialization, and where and how attention is directed.

Priority Setting

Family systems have priorities. These priorities may emerge out of family negotiation or conversation, or out of the confluence of the individual priorities of some or all of the family members, a confluence in which everyone cares enough about the same things or in which the more powerful, more vocal, more assertive family members have their way.

One place to see priority setting in action is to look at families dealing with information overload—for example, a family newly immigrated to the United States or a family newly dealing with a serious economic or health crisis. In such situations a family is likely to be confronted with far more than it could possibly deal with. Priorities must be in operation so that what is necessary for basic existence (for individuals, for the family) will typically get the most attention—e.g., putting food on the table, finding a place to live, getting a family member who is very sick to a hospital. In setting those highest priorities, other matters may never emerge from obliviousness or may drop into obliviousness. For example, the family may be oblivious to troubles their neighbors are having, or the adults in the family may be oblivious to problems a child in the family is having at school.

Priority setting is not necessarily rational. In setting survival priorities, the immigrant family, the family in economic crisis, or the family confronting a health crisis may, for example, be oblivious to resources that could help them. They may be oblivious to food and housing assistance programs or programs that can provide certain kinds of emergency assistance in a health crisis. Related to this, a family systems theory perspective on family priorities might lead us to look at a family's external boundaries. Family external boundaries are more or less open to information coming from the outside and more or less open to letting information leave the family for the outside (Montgomery & Fewer, 1988, pp. 117–18; Rosenblatt, 1994, pp. 76–77). No family can be open to everything that might come into it or can reveal all to outsiders, but when in a crisis, the boundaries of a family often become more closed to information coming in or going out. The priority is to deal with what seems to family members to be most central in the crisis. And a family that already is overloaded with crisis information will almost certainly close itself to more information coming from the outside than it would if it were not in crisis. As the boundaries close, a family in crisis narrows the range of resources they can learn about or seek, and they narrow the amount of information they will give out about their situation to others (some of whom might be able to help were they to have sufficient information). This is not to say that families in crisis necessarily make mistakes in priority setting. In setting the informational priorities they set, and in being oblivious to what they are oblivious to, a family in crisis is doing what seems to some or all of its members necessary to deal with the crisis. It is difficult for an outsider to say that they could do more. In fact, quite possibly anyone who tried to help a family in crisis by trying to overcome the family's obliviousness to this or that would not be experienced by the family as helpful but as undermining them by overwhelming them.

Family Socialization

Families always socialize members to be oblivious to a great deal. But I think it is not done by going through specific topic areas and saying, "You must be oblivious to this, and this, and this." How are family members kept away from things they are never told to keep away from? One key to the socialization of obliviousness is for the family to socialize family members to focus on what, for the family, are not matters of obliviousness. "This is what you should know. This is what you should pay attention to. This is what is important." In teaching what is important to attend to, families make other areas matters of obliviousness.

Another key to the socialization of obliviousness is the socialization of curiosity. Just as families encourage attention to some matters and thereby discourage attention to others, they also encourage curiosity about certain areas, and that is where curiosity generally focuses. For example, a family in which the powerful adults work at knowing the latest information about television and film celebrities will be socializing everyone else to be curious about those celebrities and not about other matters. In addition, I believe that families actively discourage curiosity about certain matters that are not of interest or importance in the family's scheme of things. At the extreme may be families who do not support curiosity about anything, but I think more often curiosity is shaped to stay within bounds. Curiosity about matters outside of the permissible areas will be discouraged, scoffed at, treated as troublesome, punished, devalued, or otherwise headed off. Imagine a child who is curious about what is in the gutters in the street but whose parents are only curious about the lives of sports and entertainment personalities. The parents will see no point to the child's explorations in the gutters and may see some risk in it. Whereas if the child became curious about sports or entertainment personalities, the parents probably would be engaged by that curiosity and enjoy talking with the child about those personalities.

From another angle, people learn in the family, as well as in other contexts, how to be "good." Most people want to be good. It is high praise to say of someone, "He is a good child," "She is a good wife and mother." Part of learning to be a good person is to learn to stay away from areas of shared family obliviousness. A good family member attends to what good people attend to and is not interested in much else. Being properly oblivious earns one the esteem of others in the family.

The learning involved in adapting to family patterns of obliviousness can become so well integrated into who people are that it becomes part of what could be called their personality, their basic disposition for dealing with the world. It also can be understood to be integrated into their roles in the family. As a person learns in the family to be a child, a provider, a parent, a problem solver, the entertainer, the soother, the patriarch, etc., the person learns how to carry out those roles in ways that respect family patterns of obliviousness. For example, the "good" parent does not educate children in religious thinking that has been in the family's areas of obliviousness. All this is not to say that socialization of personality and role are solely about what it is people are expected to be unaware of, but lack of awareness about various things can be seen to be a big part of personality and role.

Family socialization is, however, anything but simple. Children who arrive in a family as newborns or as adoptees or foster children are not all equally easy to socialize. Some may, for example, be inherently more curious than others. Some may find it more difficult to learn to focus on the things that the family focuses on, because other things draw their attention. As children grow older, they become more autonomous and are typically more of the time in educational, peer, and other situations where they might learn about matters that could threaten the family's shared obliviousness. Some children may continue to remember feeling forced into thinking the way the family adults do, resent it, and remember some of the thinking they were forced to give up. So children do not automatically come to fit the family's socialization pattern, and they may even come, for various reasons, to put pressure on the family to change what it is oblivious to or how obliviousness is socialized.

New family members also arrive in a family through partnering with family members who already are in the family. And that can create challenges. For example, a young adult family member brings a partner into the family whose obliviousness, curiosity, interests, realities, personality, and roles have been socialized elsewhere. Perhaps the most striking evidence of the difference will be in the first days, weeks, and months of contact of the newcomer with the family. In innocence, the newcomer may point out things that the family has steadfastly been oblivious to (for example, Grandfather obviously cheats at family card games, the television set in the family room is played so loudly that nobody can talk to one another, and the elders of the family strongly advocate an international politics of peace but cannot maintain peace with each other). The reactions of family members to the comments of the newcomer may socialize the newcomer to respect their areas of shared obliviousness, but the newcomer may also change the system by pointing out things that will forever be hard to ignore or forget in the future. Also, as with children, the newcomer may be difficult to socialize to fit the family system.

Families in the United States can ordinarily be thought of as created by the coming together of partners from two (or conceivably more) different families of origin. Each partner will bring part or all of his or her family of origin obliviousness to the new relationship. New relationships can be challenging in myriad ways, and part of the challenge may be resolving differences about what to be collectively oblivious to. He does not want to hear from her about dishonesty and hypocrisy in government, for example, and she does not want to hear from him about cholesterol in their food.

Keeping Attention Elsewhere

Families have multiple ways to keep family members' attention focused on certain matters and in the process to keep family members from focusing on or even being curious about areas of obliviousness.

Looking Away, Seeing but Not Seeing

Adults may not register or may look away from scenes that threaten their obliviousness, and in doing that they may teach their children to also look away. For example, a well-off family may drive through an impoverished neighborhood without actually noting and thinking about the neighborhood they are passing through. They may look at but not register that they are seeing decrepit housing and people inadequately dressed for the cold and windy weather. If a family wants to remain oblivious to poverty and those who suffer because of it, not registering these scenes or averting their gaze is useful in maintaining obliviousness.

Keeping Busy

Families generally have multiple, more or less compelling ways to keep their members busy. It might, for example, be housework and yard work; it might be active participation in a religious congregation; it might be putting long hours in at work; it might be busy involvement in the lives of children or grandchildren; it might be keeping up with the family's favorite television programs. All the ways of keeping busy that are part of a family's life can be said to reflect family values and investments of time and energy. They also help to maintain shared obliviousness. Putting time and energy into X and Y means there is much less time and energy available to put into A and B. Their busy involvements keep them away from what is irrelevant to their lives, but also from what could be frightening, threatening, confusing, embarrassing, or otherwise difficult, challenging, or unpleasant. It also keeps them from information overload.

When a family system keeps individual family members hard at work for long hours, one can imagine that maintaining shared obliviousness is one goal of that work. Busy activity serves, among other things, to focus attention and energy narrowly. One could say that a family that keeps very busy, for example, with adult family members working several jobs and long hours, needs the work because the income the work brings in is needed. But what is it about the family that creates the economic needs, and why those jobs and not others that might pay as much for fewer hours of work? Perhaps people have no choice about what they need or what they can earn, but if they do, there are questions to be asked that link to matters of shared obliviousness. Or consider, to take a related example, a family that works hard to meet high standards for entertaining guests. Their work at entertaining might seem to them to be about doing what needs to be done to be good to their guests. But at another level it may be a way of keeping them oblivious to certain matters by sapping their energy for perception and curiosity that might lead them to, say, challenge aspects of the family's shared obliviousness or to learn what would in some way create difficulties for the family. In general, one would want to look at who is busy and ask what busyness keeps them from.

Sometimes bereaved people say that they distance their feelings by keeping busy (there are several examples in Rosenblatt, 2000). If bereaved family members collectively keep busy, one can wonder whether the busyness is a way to keep them oblivious to something as a family. Perhaps busyness keeps them from getting to ideas that would lead to tense interactions about how a death came to be. Perhaps busyness keeps them from the awareness that they will have to collectively face difficult decisions about what to do with the possessions of the person who died. Perhaps it keeps them collectively away from talking about who in the family might die next.

Hard work can also keep a family away from questions about the society in which they live. For example, busyness might keep a family away from questioning whether the grind of work in a society dominated by corporations and wealthy and powerful elites is worth it (cf. Brosio, 1994, pp. 209–61), or whether they should find ways to be less controlled by the societal system or should try to do things to change that system. (More about this in chapter 3.)

Television Watching

The average adult in the United States says that she or he watches roughly three or four hours of television per day (Pew Internet & American Life Project, 2007). In theory, television watching could be about learning new and important information, coming to profound new realizations, and overcoming shared obliviousness. And certainly a person can learn things from watching television, but some critics of U.S. television say that it is a wasteland, that for the most part it does not inform or push viewers to think, learn, or acquire new perspectives (e.g., Rosenberg, 2004). Even news programs can be criticized as, for the most part, dealing with the same old stories in the same old ways, ignoring an enormous amount of what arguably deserves to be important news, and being controlled too closely by corporate and government officials (see chapter 3).

Also, in theory, family members could sit together while a television set was on and talk in ways that would get them to deeply connecting places with each other. And since television operation may have any of an enormous range of family uses (Alexander, 2001), it works that way some of the time for some families (Tovares, 2007). But it seems as though television often reduces family interaction because it draws attention toward what is on the television screen (e.g., Gantz, 2001; Rosenblatt, 2006, ch. 5; Tovares, 2007), and in some families, people seem to use television set operation (not even necessarily watching) to achieve communicative and emotional distance from each other (Rosenblatt & Cunningham, 1976). So the three or four hours of television watching per day of the typical individual and the greater number of total hours that a television set plays somewhere in a household may often serve to maintain shared obliviousness to what goes on within the family and to the lives of other family members.

Television watching also fills time that could be spent on other matters. By watching television, people are not learning A or B, not doing C or D, not being curious about anything else. From that perspective, too, television watching helps to maintain shared obliviousness to a great deal.

Avoiding Threatening Others

A family can more easily maintain shared obliviousness by avoiding others who might threaten that obliviousness. The family might avoid people whose realities they know or suspect to be different from theirs or whose questions might jeopardize areas of shared obliviousness. In this regard, consider families who have serious problems but who do not want to go to a family therapist. Perhaps their reluctance to go to a therapist is about not wanting the stigma they associate with seeking mental health help. Perhaps it is about feeling they can handle their problems on their own. But perhaps it is the fear that a therapist will open up areas to which the family has been oblivious, and in the process end their obliviousness. That fear concerning matters to which they have been oblivious is not of revealing secrets they know to exist. The fear is about opening up who knows what kind of vast, alarming, reality-disrupting perspectives or of revealing that underneath it all (all the work, all the television watching, all the possessions, all the daily activity, all the family conflict) there is nothing or nothing that makes any sense or something that has a very uncomfortable meaning that they have not considered.

Family members may also avoid people whose difficult situation is threatening because it reminds the family members of what might happen to them. For example, farm families who are in such serious economic difficulty that they seem likely to lose the farm may be avoided by the members of other farm families in their community (Wright & Rosenblatt, 1987). A family close to losing their farm is a reminder to other farm families of how vulnerable they may be.

Perhaps some of the time, obliviousness is maintained not by avoiding others but by avoiding the reactions of others. Consider, for example, a family that has paid for the privilege of walking to the front of a queue at an airline ticketing counter and going through a separate and quickly traversed airport security checkpoint so that they do not have the long wait of other families. Or consider a family that parks its very large luxury vehicle across three parking spaces in a crowded parking lot. Or consider a well-off family with an injured child and gold-standard health insurance at a hospital emergency room who can move past others who are not well off and move quickly into obtaining care for the injured child. Possibly a family may be perfectly aware that they are moving ahead of and disadvantaging others or that they are using more than their fair share of resources and be perfectly aware of the reactions of others. But families may also act in ways that keep them oblivious to the reactions of others to their exercise of entitlement. They may move to the front of the line while keeping their back to the others they have passed. They park the luxury vehicle and then rush inside, not staying to watch the anger of others who are looking for a parking place and see the luxury vehicle occupying three spaces. Perhaps the tipoff to the avoidances that maintain obliviousness is that they ordinarily involve an averting of gaze, quickly leaving the scene, or interactions and activities that distract the family from the sounds and sights of the anger, disapproval, and indignation of others. From this perspective, entitlement is not only about claims to precedence and an unfair share of resources. It is also about actions that maintain obliviousness to the reactions of others. "We are entitled to what we want, and we are entitled not to know about your anger, disapproval, disgust, or indignation."

Etiquette

Etiquette may serve to smooth relationships or establish that one is socially appropriate or has a certain status. Saying thank you, eating properly, or addressing others with proper respect may make it easier to get along with others, may establish that one is a proper kind of person, may locate one in a status system, and may help others to feel comfortable. When parents teach children etiquette, the manifest intent is to help the children to do better, to fit in, and to know how to make others more comfortable. But etiquette may also be in service of shared family obliviousness.

When one behaves with proper etiquette, one may be sustaining the shared obliviousness of one's family. One does not ask embarrassing questions or bring up difficult topics in the family (Nydegger & Mitteness, 1988), and that may enable shared family obliviousness to be maintained. One is not supposed to notice certain things in the family. One is not supposed to reveal certain information, impressions, feelings, or knowledge to others in the family, which helps to keep others in the family oblivious. And because one is not supposed to reveal these matters, one may stay away from knowing them and may avoid interactions with others that may bring one to greater awareness, and those proclivities help to maintain one's oblivious place in an obliviousness-sharing family system. "I won't ask about that; I won't let myself look there; I won't continue the conversation once it approaches an improper or uncomfortable area." So when we teach a child etiquette, we may be teaching the child to be ignorant and obliviousness about things in the family. Sharing obliviousness is often polite. Resisting the forces for obliviousness can be rude.

Trivia in Families

That families are of vital importance in people's lives does not mean that family interaction deals primarily with matters of great moment. In fact, often family interaction focuses a great deal on what seems by the standards of many, possibly including the family members, to be trivia, or what is sometimes called "small talk" (Coupland, 2000). "How high should the grass be?" "I saw a coffee table like ours when I visited the Smiths." "I got another unsolicited credit card application in the mail." "What is the right shade of beige to paint the bedroom?" "Should we have macaroni for supper?" "Why didn't you tell me we were running out of paper napkins?" "Mary said that her dog jumped over the fence."

A substantial amount of what family members talk about accomplishes the necessary business of living a life together. Someone has to decide what groceries to stock in the house. If a couple wants to go to bed together and bedtime is not routine, each night they have to communicate about bedtime. If one family member is going to pick another one up at school, they have to agree about when and where they will meet. But beyond the daily arranging and coordinating, connecting and informing, there may be a great deal of trivia.

There is risk in labeling anything other people talk about as trivia, because what seems to an observer to be trivia may have its uses or even deep meaning for the members of a family. Family members laughing and talking about a piece of food that has fallen on the floor may be connecting in deep and important ways. Family members may express their love and connection through discussions, jokes, stories, and the like that seem to observers to be about trivia but for the lives of the family members may be immensely important. Let's say, however, that it is possible to determine that some things are trivia and that families sometimes interact about trivia. Then a question arises about what the link is between trivia and shared family obliviousness.

One possibility is that by dealing as much as they do with trivia, family members may often together in a sense collude in staying away from what is threatening to the family system (cf. Hankiss, 2006, p. 6), and what is threatening may well include threats to areas of shared obliviousness. Related to this, quite possibly there is a systemic link between focusing on trivia and the failure to be curious. To achieve shared obliviousness, a family may have to block curiosity about much. One of the ways to do this is to socialize children and adults to focus on trivia. By focusing on trivia, they are not so often in a place to become curious about matters about which they share obliviousness.

Distraction

Families may maintain obliviousness by directing attention elsewhere whenever conversation or attention moves uncomfortably close to an area of shared obliviousness (Ancharoff, Munroe, & Fisher, 1998). The distraction may, for example, involve changing the subject, a family member doing something inappropriate that draws attention away from the area of obliviousness, or someone starting an argument about something that is distant from the area of threatened obliviousness. Any family member may instigate the distraction, but then all other family members may go along with it. So what looks like a distracting act by one family member, perhaps the family member who is most anxious about what might happen if shared obliviousness were reduced in a topic area, is actually a matter of family collusion, with other family members joining in the move to distract.

In marital conflict, partners may achieve shared obliviousness to some of their individual or shared pain and frustration by focusing on those areas of pain and frustration that they can tolerate feeling or knowing about and are willing to make a regular part of their marital conflict (Shaddock, 1998). Thus, hidden under the intensity of recurrent conflict about, say, how loud to play the television set there may be deeper issues to which they are oblivious about, for example, how one partner feels emotionally abandoned by the other or how much each really matters in the other's life.

Family Communication Rules and Patterns

As family systems theory indicates, all that goes on in a family reflects family communication rules and patterns (Ford, 1983; Rosenblatt, 1994). The rules and patterns are the map of where family interactions can and cannot go and what shapes they can and cannot take. Family members may not be able to articulate most of the rules and patterns, but they are still constrained and guided in their interactions with one another by those rules and patterns. The rules and patterns are fluid and open to change, amendment, or situational differentiation. Sometimes more than one rule or pattern is applicable, which means that sometimes family members have to deal with conflict and contradiction, perhaps by working out meta-rules about how to legislate among competing rules. That is, there may be rules and patterns concerning conflict and contradictions among rules and patterns.

From a focus on shared family obliviousness, one would expect that family rules and patterns concerning communication often provide strong "shoulds" about what family communication should deal with and ignore. The rules and patterns are enforced by the approval or disapproval of other family members. They may also be reinforced by blank stares and silence in response to certain questions or comments (Hopper, 1996). What family communication stays away from is no doubt most of the time of no importance or relevance to family members. But then they may stay away from what is frightening, threatening, discomfiting, or status-questioning for the family in general or for certain family members. They may also stay away from what is paradoxical, because paradox is confusing, or relationship boat-rocking. So communication "shoulds" may be central to the maintenance of shared family obliviousness, and the "shoulds" may arise partly from what some family members may feel threatened about even though in their shared obliviousness they may not actually know what threatens them. They do not, for example, know why they don't want to talk about the work situation and pay of people who make their clothing; they just know that remaining oblivious to the matter feels good.

They do not know why they want to avoid discussing the paradox of loving one another while not wanting to spend much time together; they just know that remaining oblivious to what that paradox means and implies feels good.

Family systems foster and maintain shared obliviousness through communication rules and patterns that construct realities and that shape reality-constructing processes. It is not simply that the rules say, "Don't go here." It is also that the rules say, "This is what is true and important" and "Those are the ways you can decide what is true and important." Included in these rules might be the idea that one must stay away from topics that are too embarrassing or painful to some family members and, in the process, keep everyone from much if any awareness about those topics. Consider, for example, the relatively small amount of communication in Japanese American families between elders who experienced internment during World War II and their younger family members (Nagata, 1990, 1991, 1998). It is possible that a very brief communication can convey an enormous amount of information; perhaps even a sigh can provide great amounts of information. But if little or nothing is said, that may produce much greater shared obliviousness than if people discussed things freely and at length. Even the elders who experienced the internment process and the camps with great pain, frustration, and awareness may be able to sustain substantial obliviousness about their experiences of internment because internment is not a topic of conversation. They may think of fewer matters, think less often about the internment, and may not have access to certain memories.

The rules of communication are not only about what is talked about or not talked about but also what can be asked and what answers can be given. A family can more easily maintain shared obliviousness about X if one of the family rules is "never ask about X." As important, but perhaps more subtle, are rules about how to ask questions and how to answer them. Families will have communication rules about how to ask questions about certain topics that push for limited answers to the questions—questions, for example, that imply or say directly, "Don't tell me more than what I want to know." Even the simple question, "How are you?" can be asked in a way that in a certain family rules environment says, "Tell me you're fine whether you are or are not" or "Only say a little bit about what is wrong, preferably in a way that minimizes information about your troubles." In accord with such rules, the system not only calls for attenuated answers to such questions but probably can be counted on to succeed in eliciting attenuated answers. And that can help to maintain broadly shared obliviousness in the family.

As part of their communication rules and patterns, families have rules and patterns dealing with privacy. For example, they have rules about interior doors being at times shut and perhaps even locked, the use of spatial or temporal distance to gain privacy for an interaction, and whispering. These rules and patterns guarantee that some in the family will be oblivious to some of what others in the family do. And related to this, families will have rules about respecting the boundaries in ways that keep some or all in the family oblivious about a great deal. For example, "We will not ever ask our extended family about physical fights in their households, and we will never listen at a door that is closed."

Family Secrets and Selective or Distorted Family History

The family systems theory literature indicates that many families have secrets that family members who know the secret keep from others in the family and from outsiders—for example, a premarital pregnancy, an abuse event, the relative who committed a crime, a mental health problem (Brown-Smith, 1998; Imber-Black, 1993). Family secrets involving some family members keeping information from others, what Karpel (1980) called "internal family secrets," may mean, for example, that Mom and Dad know that Mom had a previous marriage, but that is a secret they keep from the children. And if Mom and Dad keep the secret successfully, the children and all who become part of the children's families in the future will share obliviousness about Mom's previous marriage.

Maintaining secrets from others in the family and from outsiders can serve many uses. Sometimes a family member has a job that requires keeping some secrets (for example, a therapist or physician who must keep client/patient identity and personal information secret). Sometimes there is a sincere interest in protecting younger family members from knowledge of the horrors that older family members went through (Aarts, 1998) and protecting younger family members may protect older family members from reminders of their pain and horror (Ancharoff, Munroe, & Fisher, 1998). In fact, some secret keeping may be in service of maintaining family obliviousness, not only shared obliviousness by those in the family who do not know the secret but even by those who do know it. For example, if nobody outside the family and none of the children know that Mom had a previous marriage, Mom might be able to get through many days without thinking about that marriage. And if there is almost nobody who knows about that marriage, nobody is present to ask Mom and Dad, both of whom know about the marriage, questions that might lead to Dad knowing more than he previously did and Mom getting back in touch with things she had not thought about in many years and that are painful, embarrassing, or otherwise difficult to think about. Thus, family secrets can be partly about heading off questions that can lead to answers that break down obliviousness of the secret holders.

Another side of not asking questions is that secrets are often well preserved in families because nobody asks questions (Aarts, 1998). That is, secret keeping and lack of curiosity may often be dynamically linked. For example, the children never become curious about whether their parents had marital, romantic, or sexual relationships with others, and so it is easy for Mom and Dad to preserve secrets in those areas if they have any.

Related to family secrets, every family has a set of histories. For example, a couple has a history of their coming together and of their relationship up to now. They have histories of their separate families of origin, and if they have children, they have histories of the lives of those children. Granted that there is a certain subjectivity to what gets counted as history. Granted that there must be selectivity to history; a history cannot be about every single event that occurred on every single day. But I would argue that family histories are not selective randomly and do not necessarily exclude what is less interesting and important. I think family histories often include obliviousness to what might make certain people uncomfortable or in some way lower the family's reputation with some people—for example, the German family histories that deal with what certain family members did during the time of the Holocaust (Rosenthal, 1998). The obliviousness that is desired is partly to promote getting along better with others in the family and outside and partly to help family members to feel more comfortable. Thus, a family history may omit humble origins, unless humble origins are valued. A family history may omit extramarital and premarital affairs, perhaps family fallings-out, and perhaps actions of some family members that harmed other family members.

It may be impossible to know the unspoken history of a family, but if one knows a small amount of the unspoken history of any family it would be a bit of history about one's own family. That invites a mental experiment. Imagine being able to make known to all in your family and all outsiders the pieces of family history that you know that never get mentioned when family members speak about the family's history. Perhaps they never mention the uncle's long-term engagement and sexual relationship with a woman other than the woman he eventually married, Aunt So-and-so. Perhaps they never mention the cousin who served five years in jail on fraud charges. Perhaps they never mention the bitter jealousy between grandmother and her sisters. Once we have in mind what is omitted, the experimental question is: "Who would be upset for what reasons if we added these matters to the family history that we regularly told to family members and outsiders?" I would wager that there would be answers to the question, that it is not beyond imagining who would be upset and why. For example, Aunt So-and-so and her children might be thought to have something to lose if everyone knew that she was not uncle's first love.

Family Myths and Shared Family Obliviousness

Families have narratives or stories that they know and tell and that are often important to them (e.g., Langellier & Peterson, 2006; Rosenblatt, 2000).

The narratives construct family realities, making some realities salient, denying other realities, and ignoring (by not mentioning them) many, many other realities. One kind of narrative that has received considerable attention in family systems theory and family therapy writings is the family myth (Byng-Hall, 1973; Byng-Hall & Thompson, 1990; Glick & Kessler, 1974, pp. 30–36; Rosenthal, 1998; Wambolt & Wolin, 1988). Family myths have been defined in a number of ways, reflecting differences in, for example, whether family myths are conceptualized as individual or co-produced narratives, whether the myths are embodied in substantial and often communicated narratives or are mainly embodied in shared beliefs, and what a therapist might hear from a family in serious difficulty versus what a qualitative researcher trying to understand the lived experience of an ordinary family might hear. Here I define family myths as narratives that family members generally share, that are believed or at least not contested by family members, and that rest in part on obliviousness. The myths may be about almost anything, including the family's origins, the success of an ancestor, how a married couple in the family came together, or what a healthy family must be like. The myths may exclude, and therefore in a sense remove from memory, certain relatives, family events, and family disagreements and difficulties that do not fit the general line of thinking advanced by the myth. The myths are mythical in the sense that they are not true or not verifiable or are so selective that they leave out important contradictory information, or they exaggerate. Often, however, people are oblivious to the fact that their family myths are myths.

When all in the family are oblivious that a family myth is a myth, their obliviousness preserves a surface family harmony. And then the obliviousness reinforces whatever the myth accomplishes—perhaps hiding an embarrassing past, perhaps reinforcing certain values, perhaps justifying things that go on in the family that from some perspectives could be seen as unfair or otherwise inappropriate. The myths are, in a sense, a family's control of individuals in areas having to do with the myth (Bucher, 1985). For example, if a family's myth is that everyone in the family has always been loving to everyone else in the family, that may push family members to act loving whether or not they feel like doing so. Also, family myths can keep family members oblivious to things that disadvantage certain family members (Coleman, 1992)—for example, the myth that Mom has always been a superb cook could be a way to keep her cooking and everyone else free of responsibility for cooking.

Family members may differ in how much they accept a myth as true and how much they are oblivious to the mythic features of it. Sometimes when family members differ in whether they believe a myth or not, the obliviousness of some family members silences others. For example, let's say there is a family myth that Grandmother, who died several years ago, was sweet, affirming, and generous. The family members who repeat that myth and who are oblivious to the mythic qualities of their narratives about Grandmother may so want to believe their myth that others who remember Grandmother as nasty, critical, controlling, and even frightening may keep quiet. So sometimes obliviousness that is far from unanimous in the family may lead to what looks like family-wide obliviousness as others respect the obliviousness of some or choose not to resist it. It is possible in this situation that those who remain silent will foster obliviousness in others in the family who have no way to know that the myth is a myth. So what started out as a family myth that many in the family know to be untrue may become a myth that all younger members of the family believe is true as older family members who know there is nothing to the myth keep quiet about it and eventually die.

Family myths may be forced, shaped by, modeled by, and reinforced by myths in the larger society. For example, powerful people in government, the mainstream media, and the corporate world may work to make a war seem heroic and justified and feelings about the losses from the war primarily to be grief, thereby submerging rage and indignation that the war should not have been fought and was based on lies and that the primary impetus for the war was not heroism but mistaken and immoral political calculation (Ehrenhaus, 1993). So a family who has lost a member in the war may tell family myths about the war that include the myths coming from government, the mainstream media, and corporations. Similarly the families of Germans who were soldiers during World War II might often have the myth that their family members who were in the military were "clean" and tried to help those who the Nazis were intent on murdering (Rosenthal, 1998).

The Implicit Marriage Contract

It is possible to think of marriages as involving implicit contracts (Sager et al., 1971). The idea is that couples come together (and sometimes get into trouble) because they have implicit, unvoiced agreements that are initially attractive and satisfying to the partners and bind them into a relationship system that in important ways works for both of them. For example, "I will be well organized and strong provided you will let me have my way." Or, "You can have control of our finances and financial decisions provided you take the blame if anything goes wrong." Extending the idea of implicit marriage contract to the area of obliviousness, it is possible to imagine that couples come together and stay together in part because they implicitly agree on what to be oblivious about—for example, "We will together tune out information that makes us both feel guilty, anxious, ashamed, and otherwise uncomfortable about our comparative wealth, our smoking, or the effects of our way of living on global warming." Thus, part of the implicit contract is something like: "I can't stay with you if you pay attention to this or speak about that." Or, "I will only continue to be nice to you and go along with what you want if you never draw my attention to X."

Another way of understanding what could be called implicit marriage contracts places a couple in the context of the parental conflicts that provided the context in which each grew up, the unresolved issues between each partner in the couple and his or her parents, and the ways each partner was burdened by family expectations, hopes, worries, fears, traumas, and much else (Scharf & Scharf, 2007). The family of origin baggage each partner brings into the couple relationship can be expected to be in zones of obliviousness, but it will nonetheless play out in the couple relationships. The playing out of internal individual burdens may be in the form of conflict in the couple or of one partner externalizing anger, guilt, and other unacceptable feelings to people near and far outside the couple relationship. It may also take the form of couple sharing of hostility and prejudice toward others. And if the couple has children, the children may adopt the externalizations of their parents and be burdened by them, and the children will grow up with their own externalized expressions of the baggage they have acquired from the parenting they received. Whatever the form of externalization, these are matters for what some psychoanalytically oriented thinkers might call the social unconscious (Hopper, 1996; Scharf & Scharf, 2007) and that I would call aspects of shared couple obliviousness.

Homogamy

Many marriages are based on considerable homogamy, that is, on partner similarities on culturally important dimensions. For example, two people who share social class, ethnicity, educational backgrounds, and political values may partner. Among the ways to understand that, one can imagine that homogamy provides a good start at developing shared couple obliviousness. Sharing backgrounds and values may well mean that many areas of knowledge are given in the couple relationship, and vast numbers of other areas are relatively unlikely to be broached. In a sense homogamy is a basis for an implicit marriage contract of the form, "You and I are so alike on so many dimensions that we should be able to get along without rocking the boat for each other by opening up threatening new areas of awareness and knowledge."

Language and Family Obliviousness

Language can be a tool for maintaining shared obliviousness. One way it can do that is that it is difficult to talk about what there are no terms for. The members of some families can sustain shared obliviousness about certain matters by not having terms to talk about those matters. For example, if nobody has the terms *premenstrual tension, alcoholism,* or *white skin privilege,* it is much easier for members of a family to be oblivious to those matters. Families can maintain this kind of language-based obliviousness by avoiding conversations with others who might have appropriate terms and by avoiding information sources that might give them the words and associated concepts. In fact, it may be common that families that share obliviousness to some matter in part because they lack words to describe the matter also act to avoid learning words that could threaten the shared obliviousness.

Language can also be used in an obfuscatory way to keep family members oblivious to something. For example, if the words people use in talking about a physical fight in the family are words such as "disagreement" and "tiff," it is easier for some or all family members to be oblivious to the fact that someone in the family was physically hurt by someone else. Also, in the case of family violence, language can be used to obscure the asymmetry of the violence, making a violent encounter seem to involve all parties equally, when in fact it is one person who initiates the violence and causes most or all of the physical damage.

One could say that obliviousness is often a matter of the accidents of language, but language is not accidental. It is shaped by the same social and political forces that shape obliviousness and so is built, among other purposes, to maintain and defend obliviousness. Cohen (2001, p. xi) wrote of the challenges of involvement in an Israeli human rights group opposed to Israeli government torture. The human rights group marshaled large amounts of evidence of torture. The government, military, and mass media defenders of obliviousness to torture used language to reframe that evidence. For example, the government said that torture does not happen, the human rights group is biased, manipulated, or gullible, and, yes, something goes on but it is not torture, or if it is torture it is moral. The language builds a wall around the torture to protect it from challenges to the practice and also to guard against challenges to the widespread obliviousness to its existence. Perhaps the human rights group's challenge to the torture overcame the obliviousness of some people to the practices, but the counterattack that framed the group and the practices in ways that defended the torture could make it possible for obliviousness to return. And many Israelis had knowledge of the practices through their military service, through the military service of people they knew, and through the reports of journalists. But as Cohen (2001, p. xii) saw it, the obliviousness that was widespread in Israeli society was not that of innocents but of people who tried to look innocent by not noticing. And underneath this is a telling issue about obliviousness, that informing people about what they have been oblivious to does not necessarily end their obliviousness. Similarly in families, the naming and renaming of what has gone on can maintain obliviousness, and even providing language that should enable family members to recognize that they have been oblivious may not end the obliviousness.

Related to this, Cohen (2001, p. 1) wrote that "One common thread runs through . . . stories of denial: people, organizations, governments or whole societies are presented with information that is too disturbing, threatening, or anomalous to be fully absorbed or openly acknowledged. The information is therefore somehow repressed, disavowed, pushed aside or reinterpreted." Or the information "registers" but its meanings are "evaded, neutralized or rationalized away." In this book, I focus in part on what Cohen might call "collective denial," when whole families work at being oblivious. Cohen (2001, pp. 2–3) offered as an example of collective denial the villagers who lived from 1942 to 1945 around Mauthausen, an Austrian concentration camp where Jews and others were exterminated. Citing Horowitz's (1991) study of Mauthausen, Cohen described the lack of interest the villagers had in learning what was going on in the camp. Perhaps that could be obliviousness, but perhaps that was evasion of disclosing what they were not oblivious to but felt ashamed or horrified to confess to knowing. But let's say it was obliviousness. Then the question arises, what family, community, government, and societal processes pushed the villagers to be oblivious? The Nazis were murderers. "Why risk your life by being curious about what the Nazis did not want investigated?" No doubt awareness of the Nazi dangers was shared among family members. And no doubt people in some families made clear to one another that they did not want anybody taking risks by transmitting information that could endanger themselves and everyone else in the family (Zerubavel, 2006, p. 40). Or another possibility is that people realized that if they knew more they would be so sickened and upset that their physical and spiritual health would be threatened, so they worked at not knowing and encouraged everyone else in the family to keep from being informed or informing them.

In a larger sense, we are all embedded in discourses that construct our social relations in some ways but not others (Foucault, 1980; Hare-Mustin, 1994; Killian, 2002). Those discourses come from society, culture, and language. All discourses omit (push us toward being oblivious to) certain things, and families as they function reflect societal discourses in their patterns of obliviousness. For example, if the larger society offers a good guys-bad guys dominant discourse that breaks the world into dichotomies of good (moral, friendly, etc.) people and bad (immoral, unfriendly, etc.) people, it will be difficult for the family to see that there are more than two types of people or that goodness/badness is contextual or to see it as a matter of the standpoint from which people are viewed.

Language is also implicated in shared obliviousness in the sense that often family members fail to ask what is meant when a term that has great power but also considerable ambiguity is used in family discourse. What does

it mean to say, "That's a good school for our son," or, "I love you"? As in
the larger society, where terms such as "democracy," "guarantee," or "G-rated
film," can mean very different things from one context to another, so in families many terms could have multiple and quite different meanings. Saying,
"That's a good school for our son" may mean that one has carefully researched
the school, has great understanding of the son's capacities and needs, and
thinks that the school will provide a very good education for the son. But
it also may mean that one only knows that the school has a good reputation
with someone, that the school is affordable, that the school is conveniently
close to home, or that one of the son's friends goes there. Similarly, "I love
you" may mean one has strong and heartfelt feelings, feelings that will not
ever go away, and that make the other very special. But "I love you" may also
mean, "We are in a relationship in which it is proper to express love," "I lust
for you," "I want something from you," "I don't want you to know how angry
I feel at you," or "Let's not talk further." Based on obliviousness to the meanings of terms, people may marry; they may put themselves into relationships
that are not good for them; and they may for years be confused about why
someone who says one thing acts in seeming contradiction to those words.
Perhaps it is part of common societal shoulds not to ask someone what they
mean by a statement such as, "That's a good school for our son" or "I love
you," but obliviousness to the meanings of important terms can be a source of
considerable frustration and difficulty in families.

Religion and Shared Family Obliviousness

Religion can operate at the family level to shape and even coerce shared
family obliviousness. I believe that embedded in many religions are strong
injunctions about what to pay attention to, and those injunctions will leave a
wide span of topics for which the religion says in effect, "There is no religious
reason to pay attention to this." So the moral claims of religion can provide a
moral demand for obliviousness to matters that from that religious perspective could be seen as unimportant, inappropriate, irrelevant, or even immoral
to attend to. In fact, to the extent that religious institutions are like other
social systems, they would have their areas of obliviousness to pass on to the
families who participate in them. And a religion may even make powerful
threats to those who might want to explore an area of shared obliviousness,
for example, threatening the curious with the possibility of becoming a religious outcast or of eternal damnation.

   For many people, religion involves prayer. In shared prayer people can
collectively petition for help with what they are struggling to achieve or
what they are struggling to be like. In shared prayer they can struggle to
keep their focus on what is defined religiously as important and away from

other matters, and they can pray that their family members will also focus on what is religiously important and away from what is not. People who want to remain oblivious about this or that, who want family members to be oblivious to this or that, or who even want to restore obliviousness in self or in the family may ask God for help with obliviousness. This is not to be cynical or negative about religion. A central idea of this book is that obliviousness is a system requirement, so why wouldn't people turn to God for help in attaining the obliviousness their family system needs?

Obliviousness and Power in Families

Power happens in families; some family members get their way more than others, in general or in specific situations. Even something as mundane as who controls the television remote is an expression of power. (And in heterosexual couples the remote is often controlled by the man in the couple—Walker, 1996.) Obliviousness can be an expression of power in a system (Townley, 2006; Tuana, 2006). For example, family members with power can maintain that power by keeping others in the family (and perhaps themselves) oblivious to information that could raise questions about the legitimacy of the power or that could undermine the informational base of that power. To paraphrase Townley (2006) and Tuana (2006), obliviousness, like knowledge, is socially situated. A family patriarch, parents in relationship to children, or the adult who has more power than his or her partner may have more power to keep some things secret, to limit how others interpret things, to keep others focused some places and not others, and to silence (Zuk, 1965). In this regard, obliviousness often has a great deal to do with who has the power to define what it is people should pay attention to. Flood (1999, p. 116) wrote about "knowledge power," the power to "determine what is considered to be valid knowledge and consequently valid action." From a focus on knowledge power, a number of questions arise about families. Who in a family benefits from attention being structured this way and not that? Who benefits from deciding what it is that the children in the family do and do not need to learn in school? Presumably a person with power in the family, even one who sees self and is seen by others as benevolent, may want family obliviousness to things that would undermine that power or make the person with power uncomfortable (Karpel, 1980). So people who seem benevolent and protective of others may, among many things, protect themselves by shaping and enforcing obliviousness in self and others.

Critical systems theory (e.g., Bausch, 2001, pp. 123–37; Jackson, 1991) addresses problems arising from the unethical skewing of power in systems and approaches for correcting the skew. Among the many approaches for correcting the skew are to find ways to welcome and legitimate conflict. The skew can also be corrected by valuing multiplicity of perspectives, paradox, contradiction, dialogical processes in the system, and diversity (of experience, standpoint, and stakes) in a system. From that perspective, perhaps it is not important that a family member with power has an interest in keeping self and others oblivious about certain matters if the family allows conflict, multiple perspectives, paradox, contradiction, dialogical processes, and diversity of experience, standpoint, and stakes. Conflict, multiple perspectives, and so on will do much to counter the forces for obliviousness. From the perspective of critical systems theory, the issue in looking at power in families as it relates to obliviousness is not so much that people with power push for obliviousness as the extent to which they do so in ways that give others latitude to participate in choices about what to attend to and what to be oblivious to.

All this is not to say that people with less power are powerless. Existing gives power. Having a standpoint gives power. There is the power to think, to search, to threaten to reveal things, to leave the system, to harass the system, to reduce the efficiency of the system, to criticize the system, to question the standpoint of people who are central to the system, to be curious, to notice inconsistencies, to catch on that something is missing, to pay attention, to act covertly in opposition to system rules, to witness, to stake a claim, and so on. So what I have written here about power may apply to any family member, even a person who seems at first glance to have little or no power.

Shared Obliviousness and Family Happiness

When I Google "oblivious family" and "family obliviousness" on the Internet, many of the hits are about families who are said to be happily oblivious. Shared family obliviousness may not only be a cognitive matter, it may also be an affective matter. Shared obliviousness protects families from what could alarm or scare them, what could make them feel guilty or otherwise uncomfortable, what could challenge processes in the family, and what could overload the system. Any of those things could reduce family happiness, so shared obliviousness may often bring happiness. From that perspective, one answer to the question about why families would choose to be oblivious is another question. "Why would people not do what makes them feel happy?" If happiness is about meeting needs, finding satisfaction, and avoiding pain, it stands to reason that people would seek happiness not only through a good meal, a good night's sleep, loving relationships, and feeling safe. They would also seek it through shared family obliviousness.

From another perspective, however, some shared family obliviousness has great costs for individual family members or for the entire family. Also, the end of shared obliviousness about a matter could allow new and rewarding family connections or could end unpleasant processes of suppressing and blocking awareness. So it would seem that despite the forces in families to maintain shared obliviousness that sometimes the end of shared obliviousness about something would be welcomed by some or all family members. "It is so good that we are finally talking about what Mom went through." "Finally I understand why the family moved to this part of the world." That does not mean a family would or could end shared obliviousness in general, but eliminating a specific area of shared obliviousness may make at least some family members happy.

Conclusion

Much goes on in everyday family life that can maintain shared family obliviousness. The television watching, the conversation about trivia, the family rules about what to talk about and what not to talk about, the distractions, and so on help the system to function without information overload. The system may function more in the service of family members with power, but then everyone in the family has some power. The family system may function to keep family members (or certain family members) from feeling threatened, uncomfortable, embarrassed, or anxious or from losing status in the community. But that is not necessarily a bad thing. Families must be oblivious to most things. And also, as has been discussed briefly in this chapter and in more detail in the next chapter, it is not only the family or primarily the family that, in many instances, sets up the rules about what people in a family should be oblivious to. There are powerful forces at work in society that push for shared obliviousness to many matters.

3

Family Obliviousness to Context

A GREAT DEAL OF SHARED family obliviousness is about matters so remote as to seem from virtually any angle to be irrelevant to the family (a strike at a factory in Latvia, a rainstorm in Niger, a cow giving birth to a calf in India). However, there also is a great deal of shared family obliviousness to contextual factors in the nation and world that in some ways are actually important to the families (Hopper, 1996). Shared family obliviousness is not just about families tuning out irrelevancies in order to process important information more efficiently. It may be about the family system equivalent of ego defense in the face of threat. It may be about defending family power and privilege. It may be about avoiding family-wide feelings of guilt. It may be about maintaining comfort by avoiding knowing how much of what the family strives for and values is linked to systems designed to control, police, limit, and use the family and its members. It may be about maintaining comfort by achieving ignorance of how much other people suffer, are disadvantaged, and are oppressed. Why does shared family obliviousness to what seems like family-relevant context occur? What happens to push families toward obliviousness to contextual matters that seem quite relevant to the family?

Families Oblivious to Much World and National News

Many U.S. citizens seem often to be ignorant about what is going on in the world (Apraku, 1996; Davies, 2007; Schudson, 2000). As part of this, U.S. families arguably are often oblivious to a great deal of news internationally and nationally that seemingly could be important to them. They also could be said to be relatively oblivious to the societal processes that work to keep them oblivious to world and national news that could be important to them.

I begin the analysis of the systemic context in which families remain oblivious to much that is the news by focusing on world news. But as the analysis becomes more complicated in terms of examining additional systems contributing to the obliviousness of families I also explore family obliviousness to national news. The analysis relies heavily on critiques of corporate and politically conservative bias and control of much that is in the so-called mainstream corporate news in the United States. The critique focuses on that kind of bias because I think for decades the bias in U.S. information systems has been corporatist and conservative. But if the bias had been, say, left wing and antibusiness, I would take a pro-business, anti-Left critical stance. I think that in general the mainstream information systems serve whoever is in power, and so the obliviousness of the general public is shaped by the interests of those in power. This is a simplification, because I think every group with the capacity to get material into the news, no matter what their politics or financial resources, fosters some sort of obliviousness. But the points I want to make about the context for family obliviousness are best made in these times by critical analysis of the obliviousness creation efforts of those who have been most in power in the United States in recent years. As measured by power in the government, successful influence of legislation and government action, campaign contributions, success in the federal courts, changes in what counts as political discourse, domination of editorial pages, and control of the line of thinking on radio talk shows, I think power has been much more that of corporations and political conservatives than of anyone else.

Where does the obliviousness of so many families to much that is in the news come from? The mainstream news media do not ordinarily inform their audiences very well about what is happening outside the United States (Chomsky, 2000; Philo, 2004; Rosenberg, 2004, pp. 32–35; Thussu, 2004). Nor do general information sources such as *National Geographic Magazine* provide much information about the experiences, perceptions, identities, and realities of the people outside the United States on whom they focus (Lutz & Collins, 1993). Instead they offer materials that reinforce what Lutz and Collins (1993) describe as the ignorance, racism, colonial mentality, and cultural assumptions of their U.S. readers. They do that by emphasizing the realities of writers and editors, not the realities of the people who are being written about.

Perhaps many people in the United States do not have much interest in learning what is going on in, say, Indonesia, because they do not think much can happen there that can affect them. In fact, the general ignorance of people in the United States to what is happening elsewhere may be one with other ways in which people with power and privilege need not know what is going on because they have too much power and privilege to feel threatened by what is going on elsewhere (Schudson, 2000). But perhaps, too, at least some people avoid news from Indonesia because they suspect they would learn how the United States has harmed Indonesians, and such news could be very uncomfortable for them. If a U.S. family is unaware of, say, the ways the U.S. government may have undermined Indonesian democracy, supported Indonesian government corruption, acted in ways that have been harmful to millions of Indonesian families, or been responsible for many deaths in Indonesia, nobody in the family is made to feel uncomfortable. If family members become knowledgeable about how the United States has harmed Indonesia, there is risk of family discomfort and feelings of guilt. In fact, the U.S. government, by many accounts, supported a military coup in Indonesia that was responsible for thousands of deaths (United States House of Representatives, 1977, pp. 72–73) and subsequently supported a military dictatorship in Indonesia that was responsible for hundreds of thousands of deaths, particularly in East Timor (Nevins, 2005).

There are similar stories of U.S. government responsibility for many deaths and for undermining democracy in many other countries (e.g. Farah, 1999; Kinzer, 2006; Kornbluh, 2003). Information is available about these murderous, alarming actions of the U.S. government, and sometimes some of the first or most revealing analyses have been in the mainstream media. However, other mainstream outlets do not necessarily pick up such groundbreaking or revealing stories or follow them up. I think much more of the news about the harm the U.S. government has done to other countries is available from the alternative press, alternative television such as Link TV, Free Speech TV, and Democracy Now, and increasingly from bloggers and other Internet sources. And many individuals and families in the United States turn to the alternative press, bloggers, and other sources alternative to the mainstream media for information. But with the mainstream media saying little, and with most people in the United States not paying attention to the alternative news sources, it is easy for most U.S. families to be oblivious to the dark role of the U.S. government around the world.

In a sense there is an unspoken, unwritten agreement between many U.S. families and the U.S. government concerning what the U.S. government does internationally, an agreement that requires these U.S. families to be oblivious to a great deal. From the viewpoint of most government leaders (of both major parties), the agreement might resemble: "We'll do what we think is necessary to further U.S. interests, and it would be best if you didn't know what we did and assumed that all we did was benign." Perhaps there is some rationality to that, in the sense that it would be difficult to run even an organization of modest size, let alone an enormous government, if all stakeholders monitored and offered input into all decisions (Schneider, 2006).

From the family viewpoint the agreement with the U.S. government might resemble: "You do whatever it takes for us to continue to live comfortably. Let us have our automobiles, imported fruit year around, no worries about moral responsibility regarding people in other countries, clothing and electronic goods made in countries with cheap labor so our costs are reduced, and social order in public spaces, and we won't ask questions, criticize, pay attention, or do anything that might get in your way." Symptomatic of the unspoken agreement, government officials often do not inform the public or want the public to know what the government does internationally (e.g., Kinzer, 2006, pp. 4–5) or even domestically (Retsinas, 2007; Salam, 2001). In fact, the U.S. government uses government secrecy not only for security reasons but to keep U.S. citizens from knowing what the government does (Bianculli, 2000, pp. 197–244; Lacey & Longman, 1997, p. 65; Mitchell & Schoeffel, 2002, pp. 10–12). The executive branch even uses secrecy to keep members of Congress from knowing what the executive branch does (e.g., Scahill, 2007). And both the executive and legislative branches of government also use lies to manipulate public opinion (Lacey & Longman, 1997, pp. 65–66). So a considerable amount of what the government does may be covert, both in the sense of hiding what it does from people in other countries (Kinzer, 2006, pp. 109–216) and also of hiding it from U.S. citizens (Salam, 2001).

Perhaps government secrecy would be more difficult to maintain if the public demanded information from the government (for example, about the United States' part in wars around the world and in changing, manipulating, or shoring up governments in other countries, or about the covert means the government uses to prevent criticism from within the United States) and demanded that the mainstream media report much more fully and comprehensively about the government. But the public does not, for the most part, seem to want to be informed more than it is. In fact, some polling data show that many in the public may not want to be informed at all about some matters. For example, a Pew Research Center poll of August 8, 2006, showed that 50 percent of those polled thought that it had hurt the interests of American citizens (as opposed to 34 percent who thought that it had helped the interests of American citizens) that news organizations had reported that the federal government had been secretly examining the bank records of American citizens (retrieved from Polling the Nations Web site, January 3, 2007). And the news report about government examination of banking records was a rare breach in the great wall of secrecy around a great deal of what the government does. So it is no surprise that the public is often said to be overwhelmingly ignorant of what the government does (Bennett, 2003; Weinshall, 2003).

With obliviousness to how actions of the U.S. government frequently create, maintain, and defend certain privileges for many U.S. families, these families can feel that their privileges are deserved and can be taken for granted. They do not have to feel that their privileges were obtained through the use of force and threats, the undermining of elections and elected governments in other countries, bribery, economic bullying, domestic spying, and other processes that by some standards would be considered immoral. They can avoid feeling guilty and avoid concerns about the desirability of changes in how the U.S. government participates in the world that would decrease their privilege.

If these speculations about U.S. families are correct, I would expect that if a family member became active in learning about international matters from sources that report how the U.S. government has caused harm in other countries, that family member's learning could be threatening and uncomfortable for other family members. They might criticize and marginalize the learner. They might not like the learner telling them much about what she or he had learned, because such information could potentially be threatening and uncomfortable. It would contradict the "everything is okay" sense they get from being oblivious and from focusing on obliviousness-reinforcing mainstream media. In defense of their obliviousness, they might put pressure on the adventurous family member to keep her or his learning about international matters superficial, focused on benign mainstream media reports or on less threatening topics such as music and cuisine.

The Mainstream Media and the U.S. Government

The obliviousness many U.S. citizens have to what is going on internationally and to the U.S. government's role in what goes on internationally is facilitated by a mainstream U.S. media that reports nothing or almost nothing about many important stories (Chomsky, 2000; Mitchell & Schoeffel, 2002, pp. 13–15; Parenti, 1997). The U.S. mainstream media both manipulate the public and often allow the government to manipulate them (Bennett, Lawrence, & Livingston, 2007), and they also allow corporate manipulation. Allowing manipulation thus makes the mainstream media manipulative. The manipulation by the mainstream media takes a number of forms. When there are accusations of wrongs committed by the U.S. government, U.S. corporations, or clients of these entities, obliviousness is often promoted in the mainstream media by reporting in a "balanced" way (Jensen, 2006; Pedelty, 1995). The balancing gives propagandists for apparent perpetrators of harm as much respectful coverage as their accusers, and the balance is enacted even when reporters in the field know that the propagandists are lying. Lies and distortions are typically treated with as much respect as direct evidence that shows the lies and distortions to be lies and distortions. For example, the accusers show photos of hundreds of slaughtered women and children; the defenders of what was done say "nothing happened; no one was killed"; and the mainstream media typically conclude that the stories are contradictory. As a consequence of the balancing, the media often mask how much harm the U.S. government (and U.S. corporations and governments supported by the United States) do.

The mainstream media often, with the help of government and corporate information sources, set up what can be considered to be false or narrow debates in which the contending positions in the debate both accept a number of perspective-limiting assumptions as true (Chomsky, 2008). As a consequence, the appearance of democratic debate and an examination of alternatives pushes for obliviousness to the ways that the underlying, taken for granted assumptions could be questioned. For example, a debate over whether Iran has interfered with the U.S. forces in Iraq and should be punished might juxtapose those who say, "Yes, retaliate" against those who say, "Maybe they didn't interfere with U.S. forces, so we should wait until more information is in" (Chomsky, 2008). Chomsky (2008) points out that hidden from sight in that debate is the underlying assumption that the United States owns the world and has the right to be wherever it chooses and to defend itself against any country that intrudes into whatever the United States owns. More generally, says Chomsky (2008), to maintain the appearance of democracy there must be the appearance of debate over issues, but the mainstream media and others in power can see to it that the debate does not probe into the assumptions that underlie the media, corporate, and government power and privilege.

The U.S. government has spent millions of dollars to plant stories in the U.S. mainstream media (and also in the foreign press from which the stories could migrate to the U.S. mainstream media) in order to sway U.S. public opinion (Berkowitz, 2007). The U.S. government seems, at least in Iraq, to have been responsible for the death or detention of quite a few journalists whose reporting might have revealed matters the government does not want revealed (Davies, 2007). So obliviousness about what the U.S. government does is fostered by (1) false, biased, and misleading stories that the government directly and indirectly feeds to the mainstream media and (2) the government's work at silencing independent reporting.

One could say that the mainstream media must, like other systems, be oblivious to much. There is too much to investigate or report. But arguably the mainstream media in the United States often fail to report much, if anything, about what seem to be very important stories (Chomsky, 2000; Jensen, 2006). The mainstream media accept government censorship on certain matters (e.g., on visual reporting of U.S. casualties in Iraq—Arnow, 2007). The mainstream media even underreport stories about the ways the news has been distorted or censored, for example, underreporting information about the U.S. government creating false news stories to manipulate public opinion

(Burton & Fansetta, 2006; Hollar, Jackson, & Goldstein, 2006). Why do the mainstream media so often limit what is reported? One answer is that most mainstream media are for-profit businesses, with financial interests in limiting how much the public makes trouble for corporations like them (Hollar, Jackson, & Goldstein, 2006; Mitchell & Schoeffel, 2002, pp. 13–15; Pedelty, 1995). And then the managers and owners of major media outlets belong to the same privileged elites that run other corporations or that have powerful positions in government (Chomsky, 1989, p. 8; Manning, 2001, pp. 82–84), and so for them there is a great deal of identity of interest with the other national elites in limiting what the public knows that could threaten the interests of these elites. But then too, families might generally not want to know what is threatening or uninteresting, and so their preferences about watching, reading, and listening shape what the media provide (Stacks, 2003/04). Furthermore, families do not necessarily want to be jarred out of obliviousness to events and policies that, if they knew them, would make them uncomfortable. And then government leaders do not want to be embarrassed, criticized, or caught in immoral or illegal action by reports in the news (Mitchell & Schoeffel, 2002, pp. 10–12); they do not want to lose their power, reputation, or jobs and do not want to let down their elite colleagues and sponsors. And so those in government are careful about what gets into the news. In addition, the media cannot so easily keep their government sources, their tax breaks, their low cost use of the airwaves, and their effectiveness in lobbying government for what they want if they antagonize powerful people in government (Chomsky, 1989, p. 9). So the media are careful about what they report about government, which may explain, for example, how it came to be that the news media were virtually silent for years about the activities of the U.S. government and the clients of the government in Afghanistan (Kolhatkar & Ingalls, 2006) and Central America (Mitchell & Schoeffel, 2002, pp. 10–12). In fact, reporters and editors who do not understand what their employers want, what government sources want, and what the general public wants and who publish what is embarrassing to the government can be chastised, demoted, lose their jobs, and perhaps be driven out of journalism (Ireland, 2006).

In racial matters, too, the major media outlets have often been slow (even by decades) to report the stories of racist oppression in the United States or have presented stories in ways that keep their readers, viewers, and listeners ill informed about how the racial/racist system in the United States operates (Roberts & Klibanoff, 2006). During the civil rights era, for example, television news generally reported on dramatic protest events by blacks without offering the viewing audience anything like sufficient understanding of the problems underlying and motivating the protests (Roberts & Klibanoff, 2006). So in a sense the mainstream media were manufacturing attitudes toward black protesters while maintaining white obliviousness to white racism (Roberts & Klibanoff, 2006). Eventually, the mainstream media reported on enough instances of blatant discrimination and racism and enough instances of police brutality in the South to help get civil rights legislation passed. But by not allowing much of a voice to African Americans in the South and the North they also helped to maintain a national racial system in which white people continue to have skin privilege and to be comfortable with it by maintaining white obliviousness to that system and to their privilege.

Educational Institutions

Another party to what seems to be a kind of collusion involving families, government, and the mainstream media is education. Formal educational institutions must of necessity promote obliviousness to much. There is too much to teach. But why do educational institutions ignore what they ignore? Arguably, formal education in the United States generally trains people to take on the roles that the U.S. version of capitalism needs to have filled, but that curriculum foundation may be generally hidden from most people (Chomsky, 1996, p. 31, 2000; Margolis, Soldatenko, Acker, & Gair, 2001). The people who have the most control of the system and benefit the most from the system need others to fit the roles the system must fill and not to object to the education or try to change the larger societal system (Chomsky, 2000). So education pushes for obliviousness in many ways. It strongly reinforces certain values at the expense of others, so certain ideas, perspectives, and ways of thinking are dealt with superficially or not at all (Chomsky, 2000; Hill, 2005). Hence, the choice of what to teach about history supports perspectives and values that reinforce certain political and ideological positions, and the history that is taught is oblivious to information that would challenge those positions (Chambers, 2001, pp. 35–36). The educational system pushes for obliviousness at the family level by how it justifies the curriculum offered, how it is antagonistic to criticism and questioning, and how it justifies education in terms of jobs. It also pushes for family obliviousness through curriculum choices about what history and current events to teach and through choices about how much and how well to teach foreign languages. Even school dress codes push children and their families toward producing docile citizens who fit the roles that need to be filled in society (Raby, 2004). And the education system prepares people in journalism (Jensen, 2006), government, education, and throughout society to accept and not question the societal system, including the areas of obliviousness that are central to the system.

It has been said that public K-12 education in the United States generally ignores international matters and the role of the United States in influencing, if not controlling, the economies, governments, wealth distribution, and politics of many countries (Christian, Pufahl, & Rhodes, 2005; Zhao, 2006). One could argue that there is a collusive relationship with families, the government, and the mainstream media concerning coverage of international matters. Similarly, there is a collusive relationship concerning coverage of national news and critical analysis of the institutions that provide the context of obliviousness for families. And families do not want their children learning what the families would prefer to be oblivious to. Public education in the United States even tends to stay away from the difficulties that may occur in the family context. And by staying away from those difficulties, the schools become part of the silencing and normalizing of certain family difficulties—for example, domestic violence (Hall, 2000). And then there is government's role in shaping education. Government, which provides considerable financial support to much of education, does not want educators to teach what government leaders prefer that people be oblivious to. And the mainstream media could find itself in a difficult position if education kept pushing into areas to which the media have been oblivious. If educators did that, they would be criticizing the media and in a sense devaluing and lowering demand for what the mainstream media have to offer.

Corporations

Another system involved in what could be taken as colluding with families, the mainstream media, government, and education in the production of obliviousness is the corporate world. Corporations may be diverse in how much they work at keeping people oblivious, but it is clear that many corporations try to keep people oblivious to certain matters (Gelbspan, 2004, pp. 39–40, 47–56; Yang, 2004). They may sponsor "scientific" research and advertising campaigns that keep the public from realizing that an enormous amount of evidence shows that a corporate product is harmful (Gelbspan, 2004, p. 40; Proctor, 1995, p. 102). They may work with their allies in government to keep the public ignorant by suppressing certain kinds of information or providing disinformation (Gelbspan, 2004, pp. 37–61). Some corporations and industries have invested a great deal of money in raising doubt about the existence, causes, and possible consequences of global warming, attempting to neutralize a cumulatively enormous scientific literature (Gelbspan, 2004). Corporations responsible for illnesses and deaths, for example, through toxic material released into sources of drinking water, invest considerable money in constructing a view that makes it seem that what they do is benign or even beneficial (Carey, 1997).

If sponsorship of corporate-friendly research and skilled public relations efforts are not sufficient, corporations can use the threat of lawsuits to bully journalists and media outlets into not reporting on certain matters or into reporting that is much more favorable to the corporation than honest (Soley, 2002; Yang, 2004, addressing reporting on the environment). Lawsuits and the threats of lawsuits are also used to silence neighborhood groups, environmentalists, consumer watchdogs, good government groups, and others whose accusations, testimony, and complaints might create bad press for a corporation (Pring & Canan, 1996, March 29; Quick, 2006, April 3; Soley, 2002; St. Louis *Post-Dispatch*, 2004, September 24). Then too, corporations can use their advertising dollars to shape and limit what is reported in news sources that rely on advertising dollars (Manning, 2001, p. 82; Soley, 2002; Stacks, 2003/04); the mainstream media are careful not to offend advertisers (Chomsky, 1989, p. 8; Jensen, 2006). More generally, many corporations invest substantially in public relations, which can be considered their way of keeping the public from understanding what corporate leaders want and do that would upset the public if not understood the ways corporate public relations "spin" things (Chomsky, 1996, pp. 15–17).

Mass Forgetting

And then there is what might be mass forgetting. For example, it seems that many people in the United States have forgotten ("Vietnamnesia") aspects of the Vietnam War (Beattie, 1998, pp. 27–34). The forgetting includes not paying attention to information about the Vietnam War, remembering little, and recasting what is remembered in a way that simplifies and distorts what happened and reduces the emotional pain in the United States. Presumably, many families get caught up in this forgetting with help from the mainstream media, educational institutions, corporate efforts to focus attention, and particularly what people in the government do to focus attention (Beattie, 1998, p. 28). As Beattie (1998) saw it, Vietnamnesia is a strange sort of healing in that it involves sliding away from the moral, political, and social issues related to that war, so one could say Vietnamnesia is a shared motivated forgetting.

Why the Collusion?

Family obliviousness in the United States to what the national government does to people in countries around the world may not be simply a matter of family information management in order to avoid information overload. It may also be about protecting family privilege by remaining ignorant about what could cause family members to feel guilt, distress, or other kinds of discomfort. It also may be about a seemingly collusive relationship between the government and families, with the mainstream media, education, and corporations also joining in the collusion. One could say that family obliviousness in these situations is patriotic and conformist. One could also say it is about allowing those with great power to use that power, including vast military power and the many deaths it produces. Among the central underlying reasons for the seemingly collusive relationships that leave U.S. families so oblivious is that the relationships enable governments to achieve certain domestic and international political goals, many of which are pleasing to corporate leaders. And then the collusion would also enable many families to achieve certain comfort levels, ways of dealing with contradictions between values and events, ways of avoiding feelings of guilt, and economic goals.

Families Attuned to Alternative Media

Some families who pay attention to the alternative media seem not to share the obliviousness of most families to certain aspects of what is going on in the world. That does not mean that they are lacking in obliviousness to much that goes on in the world, but they have managed to escape some of the constraints of a corporate press, a government that wants to get what it wants, the obliviousness of many in the country to the ways the U.S. government may have harmed others in the world, and what advertisers would rather consumers did not know. No doubt the world is better off because there are families attuned to the alternative media, but I imagine such families pay costs in that in exchange for avoiding obliviousness to what others tune out they may seem to the others to be ignorant and biased. That is, if a family's areas of obliviousness are not congruent with those of families around them, there could be costs to pay. Furthermore, it may be that once a family gives up the secure obliviousness that goes with relying on the mainstream press and trusting the government, there may be a permanent insecurity about the adequacy of what the family members think they know. A critical view of mainstream information sources may lead to a critical view of all information sources.

Families Oblivious to How They Are Shaped and Limited

In systems that are organized hierarchically, patterns tend to replicate down the hierarchy (Rosenblatt, 1994, p. 68). So if a larger system that is an ecosystem for the family (government, for example) has a pattern of frequently using violence and deception, smaller systems in that ecosystem, including families, are also drawn into patterns of violence and deception. So just as governments, education, and the mainstream media work at achieving obliviousness and at keeping others oblivious, so may families.

Another way to think about the linkage of families to larger systems is to see families as shaped, limited, and used by larger systems (Donzelot, 1979; Salam, 2001). To accomplish this, the larger systems often need and can generally rely on families to be oblivious. There may be very substantial social

class differences—for example, with working-class and middle-class families oblivious to how they are shaped to become the workers and consumers
the society needs and not to protest or question the system that shapes them
(Donzelot, 1979; Salam, 2001). From this perspective, families are shaped to
want what they want, strive for what they strive for, tolerate what they tolerate, and ignore what they ignore in service to larger systems that benefit rich
and powerful people and corporations (cf. Brosio, 1994). For example, middle-class families can be understood to strive for education for their children,
a home, status, and the good life by their standards in ways that keep them
policed, limited, and engaged in activities that sustain the larger system. In
Donzelot's (1979) thinking, the policing of families, which is facilitated by
various social institutions, including education, medicine, psychoanalysis,
and social work, keeps them working in the service of capital. So the family's
very sense of happiness is bound up in what serves capital (Brosio, 1994, pp.
209–61), and that service is sustained by family obliviousness to how their
ideas of happiness are controlled and limited. If they were not oblivious, they
would presumably resist the system and live with alternative ideals and goals,
ones that would take power away from the policing institutions and restore to
families the ability to live a very different life (Donzelot, 1979).

An interesting perspective on the Donzelot (1979) line of thinking about
obliviousness is to consider how the policing institutions such as education,
medicine, psychoanalysis, and social work intersect with families. Imagine a
family in which somebody has become critical of the family way of life and
goals and claims that a way of living that is not in service to corporations, the
rich, and the powerful is most desirable. Perhaps the person wants and advo-
cates a life of minimal reliance on material goods, education outside of the
tracks available through formal educational institutions, and a life in which
paid employment is avoided. Other members of the family, who are embedded in the system as Donzelot (1979) saw it, might well refer the person to
policing institutions. Rather than seeing the person's view of what to do as a
reasonable alternative set of values, they might see it as a mental health problem or one of insufficient education. So they might make considerable effort
to persuade the person to seek therapy or education. They might even try to
commit the person to a psychiatric facility.

Families may also be oblivious (and even work at maintaining obliviousness) to how they are shaped by societal shoulds, representations, stereotypes,
and restraints concerning gender and sexual orientation. Being a proper
woman or man in the family by societal standards may place serious constraints on how family members relate to one another, resolve family disputes,
solve problems, get to know one another, see fairness in household division of
labor, and are sexual. And at the same time the family may experience how
gender and sexuality are done in the family as coming out of the essential

nature of human biology. They are likely to be oblivious to the forces in society that push and limit them in terms of gender and sexuality. They are likely to be oblivious to how their notion of the essential nature of the genders has a history, that patriarchal and heterosexist forces, rather than human nature, created those notions (Chambers, 2001, pp. 42–48). Similarly, families may be oblivious to the history of ideas of what proper parenting is (Chambers, 2001, pp. 52–55), taking their ideas of parenting as culturally reasonable and perhaps scientifically verified, and yet these ideas may be understood to serve the corporate world, patriarchy, and other forces in society and not actually to be essential for child and family well-being.

Feminist thought and action have fought the obliviousness families have to the forces that shape and constrain gender, to open up individuals and families to awareness and knowledge that could enable them to resist the societal forces, and to achieve fairness, justice, and fulfillment. Similarly, those who have spoken out against homophobia, racism, ageism, ableism, ethnocentrism, and other oppressive ideologies have inspired many to try to overcome family obliviousness to forces in the larger society that shape and limit how the family functions internally and relates to those outside the family.

Family Obliviousness to the Traumas of Others

Families are oblivious to most of the trauma that others experience. Even if we know national and international statistics about rape, homelessness, family physical and sexual abuse, police brutality, torture, and victims of natural disasters, we rarely hear a word from victims about their experiences. And even when there is a news report in which a victim has a voice, it is only a very brief snapshot of what probably would take a lot of words to express adequately.

One could blame the news media for negligent reporting, but it also may be that there is not much demand for reports of trauma. Why might families not be interested in learning a great deal about other people's experience of trauma? Perhaps that is a silly question, because it seems that most of those traumas are irrelevant to any specific family and because in a world in which families must be very selective about information coming in they will have to be oblivious to most instances of anything. But I would argue that there is more to obliviousness to trauma than avoiding information overload, that it is also about avoiding threat to the family status quo. If family members were aware, in ways that were meaningful to them, of how many people need help desperately or how vulnerable the family is to various forms of brutality and disaster, family members might become very uncomfortable and might find it difficult to continue along the everyday paths they had been traveling. Then too there is guilt to avoid about not offering help, or more than token help, to those who desperately need it. And so obliviousness to most or all trau-mas allows family members to avoid feeling that they are not good or moral people as well as that they are vulnerable.

If I am right about the avoidance of feelings of vulnerability, there might be more obliviousness to the trauma of others by families who might find the trauma most threatening. For example, people living in a floodplain might be least inclined to learn about flood victims. And then if I am right about avoiding feelings of guilt, it might be those who are most able to provide help to needy others (for example, very wealthy families) who would be most likely to be oblivious to the existence and needs of those needy others. That implies that families who seem oblivious must have some way of sensing what is out there to know and of choosing to be oblivious to those whose situation might make them most uncomfortable (more on this in chapter 9).

It could be very telling to see what happened in a family if one family member became attentive to the traumas of others. For example, let's say that one family member started working in a battered women's shelter or became a reporter in a war zone. A case can be made that because the family member's learning about the trauma of others threatened family obliviousness that oth-ers in the family would encourage the family member to find another occupa-tion and would not want to know the details of what the person was learning. Their stated motivation might be about safety or peace of mind for the person learning about the traumas, but there is likely to be a part of it that has to do with protecting family obliviousness.

At an extreme in desiring obliviousness to the trauma of others might be those families who benefit directly from the trauma. For example, if a family benefits financially from stock in a mining company that is responsible for suffering by people displaced by the mining and for deaths from poisoning by mine wastes, that family may work very hard not to know about the losses, grief, rage, and trauma of those who were displaced and the deaths of those who were poisoned because of the mining.

Whiteness and the Obliviousness of White Families

Many who study race, racism, and whiteness say that white people in the United States typically are unaware of the unearned privilege, benefits, and advantage their skin color brings them (e.g., Dalmage, 2004; hooks, 1992; McIntosh, 1988; Rothenberg, 2000; Sullivan, 2003, 2006). They ordinar-ily are unaware of what they do to maintain the system of white privilege (Karis, 2006; Sullivan, 2006) and of the arbitrariness of the system, its cat-egorizations, and its injustices (Martinot, 2003, p. 24). They are also oblivi-ous to how the privilege system and racism create their identity, and that identity includes a role in maintaining the privilege system (Martinot, 2003, pp. 197–98). Even obliviousness itself is a privilege (Martinot, 2003, p. 197). To accomplish this obliviousness may in some ways be effortless, since white people typically function in an environment of other white people in which white privilege is so taken for granted that it is easy either to be unaware of the privilege or to assume that the privilege is unremarkable, proper, appropriate, and deserved (McIntosh, 1988; Sullivan, 2006). And education at all levels may reinforce for white students an ignorance of the racial system and their place in it (Outlaw, 2007). Yet, despite white obliviousness, whiteness is a standpoint taken by white people for viewing themselves and others (Frankenberg, 1993; Karis, 2004). To accomplish the obliviousness and to live with the paradox of tuning out whiteness while using it for purposes of privilege and standpoint, one might assume that many white people must in some ways deceive themselves (Dalmage, 2004; hooks, 1992; Karis, 2006; Ortega, 2006; Rothenberg, 2000; Sullivan, 2003, 2006; Tuana, 2006; Yancy, 2004).

With so much involved in being white and being oblivious to one's whiteness, white people have to learn how to be white (Conley, 2001; Moon, 1999). They have to learn to be unaware of their privilege, to ignore evidence that they are privileged, to ignore the voices of people of color who speak critically about white privilege, and to be unaware of what they do (for example, not being curious about the lives of people who are nonwhite) that helps them not to feel guilty or to realize that the system works to benefit them (hooks, 1992; Ortega, 2006; Tuana, 2006). They have to learn these things even while working actively to gain the benefits of privilege. For example, a white mother working to develop social networks for her child chooses a "good" school for her child, and picks the right kinds of after-school activities (Byrne, 2006, writing about mothers in Great Britain). She almost certainly realizes that her choices will give the child a better future by certain standards, but she can remain oblivious to how the means to that better future and the better future itself involve building on and using white privilege.

How do white people learn to deceive themselves? And to the extent that they are unaware of their self-deception, how do they learn to be unaware? Arguably, much of the necessary learning happens in and is fostered by white families (Moon, 1999; Tatum, 1997). Moon (1999, citing Rich, 1979) and Tatum (1997) wrote about white people in the United States learning in the family to be disconnected from whiteness, evade whiteness as an issue, and to see issues of race, racism, and the power relations between the races as irrelevant to themselves. At the same time, Moon wrote, whites learn to see whiteness as normative, as the proper condition for humans, and deviations from that as what is remarkable and worthy of attention. So whites typically learn in the family to avoid looking at whiteness, but instead to look at those who are nonwhite, to understand the racial system and differences between races as about what nonwhites do and are.

To the extent that white families in the United States are able to accomplish what Moon believes many of them accomplish, many white people will learn in the family what not to be curious about concerning race and racism. For many, that will mean being uncurious about white privilege, acts of racial injustice, and most of the history of race in the United States. White people will even be incurious, and largely oblivious, to how the U.S. government has acted to make life better for white people and worse for black people (Lipsitz, 2006, pp. 1–23). Instead, they would learn to ignore the writings, the voices, and the events around them that might threaten their view that the world is fair and that they and their families have earned and deserve what they have. They would avoid learning that they have benefited from an unfair system. They would learn not to see how what they believe and do supports the system of privilege that causes terrible disadvantage and pain to many who are not white. For example, they might learn to think that "color blindness" and thinking "we are basically all the same" opposes racism, even though what it also does is to deny the enormous differences between their own experiences and the experiences of people of color. Or, to take another example, they might learn not to see their interest in feeling comfortable and safe in a multiracial situation as connected with the unearned privilege of whiteness (Dalmage, 2004).

White people in the United States learn to think about what they have (their jobs, their housing and other assets) as gained through effort and hard-earned success, tuning out the larger system that gives their effort more rewards than are given to similar efforts of people of color (Karis, 2006). In a sense, they have to learn not to see and think. But at that same time, according to Moon (1999), they learn in the family setting to be good people, so the learning they do is not simply neutral, it is learning to be "good" and respectable by the standards of their (white) family and (white) community. And implied in this is that without the respectability of whiteness, a person (well, a woman—Moon wrote about women) would lack credibility and standing in the white community. Related to this, quite possibly a white family loses credibility and standing in the white community if one of their offspring deviates from the standards of whiteness. This may be why at least some members of some white families strongly oppose an offspring's partnering with someone who is black, even to the point of "disowning" the offspring (Dalmage, 2004; Rosenblatt, Karis, & Powell, 1995). The family may be oblivious to its skin privilege and to the racial system that disadvantages black people in so many ways, but it is not oblivious to the status implications of a white person partnering with someone who is black. And then with obliviousness to their skin privilege and to the harm caused black people by the racial system white people have to find some explanation for how it comes to be that, on the average, black people do less well economically and in other ways in the United

States. Small wonder, then, that opinion polls show many white people blaming the problems that black people have on the failings of black people and not on the system that advantages white people at the expense of black people (Lipsitz, 2006, p. 19).

Presumably these aspects of white family life have existed in the United States for generations. So in white families that have been in the country for generations, there has been intergenerational transmission of obliviousness to the family's skin privilege. White people learn to think of themselves as good people, and to be good people they have to learn to distance information that would lead them to think they benefit from and perpetuate a racist system (Sullivan, 2006). And it is easier to function in this way if people have learned not to discuss "their own social, political, economic, and cultural investments in whiteness" (Yancy, 2004, p. 4).

Still another part of the socialization of whiteness is that in the event a white person is able to perceive aspects of the system of race privilege, he or she is likely not to protest it or try to change it (Yancy, 2004). If, for example, the cashier at a supermarket asks a black customer for multiple picture identification cards but does not ask a white customer, and the white customer is aware of the differential treatment, does the white customer pull out multiple picture identification cards when it is her or his turn to pay, question the clerk about the double standards, or complain to the manager? Moon and others suggest that many white people have been well socialized not to rock the boat of white privilege by challenging or questioning. Not rocking the boat is learned in and reinforced by family, and a big part of it is to learn to be oblivious to discrimination and differences in privilege. Another part is that if a white person sees a racist occurrence, she or he sees it as a problem for others, perhaps especially the targets of the racist occurrence, to deal with and resolve.

Still another way of socializing whiteness to maintain the white advantage is for white adults to teach their children to be oblivious to whiteness. So if issues of race come up, white people would then focus on people of color. It is the focus on, say, a frightening black male that keeps white people from focusing on themselves as frightening white people who are unfairly advantaged (Dalmage, 2004; hooks, 1992; Yancy, 2004). Then, if a white family socializes children to be oblivious to whiteness and racial privilege, that sets off needs for still other kinds of obliviousness. If one is socialized to be oblivious to whiteness and privilege, one probably must be socialized to be oblivious to the socialization of obliviousness. Even socializing a child regarding ideas of justice, fairness, and equality may have to be done in a way that keeps a white child and family oblivious to their part in a system that is unjust, unfair, and unequal. "Be a good girl" probably presupposes a system in which people who benefit from racial advantage find ways not to call themselves racist. One way they find not to call themselves racist is that whites have come to see specific acts that attack blacks as racist but not to see complicity (whether aware or unaware) in a racist system as racist (Lipsitz, 2006, pp. 20–21). Thus, burning a cross on a black neighbor's lawn is seen as racist. But living in a neighborhood where everyone is white and is living there because they (and perhaps their parents and grandparents) have benefited substantially from a system of white advantage that includes government benefiting whites (for example, through housing laws and how those laws are applied, social security laws, government-sponsored urban renewal, and tax laws) is not seen as racist.

Then too, beginning even in childhood a white person might have the power not to be questioned about obliviousness about white privilege and the experiences of people of color. It is people with power whose perspectives and practices will be ignored and unexamined (Taylor, 2004). However, Frankenberg (2001) pointed out that whiteness is not always ignored, that in the history of imperialism, capitalism, and race there have been times when "white" people were very aware and self-conscious of themselves as a category of people (cf. Thandeka, 2002). Sometimes that awareness went with a sense that as whites they deserved the spoils of conquest and dominance. And at times white people acted to drive black people away from certain resources, perhaps breaking the law in doing so, and perhaps getting away with their criminal acts because the white-dominated police and courts did not act to curb them or punish them (Rubinowitz & Perry, 2001/2002). And in South Africa, where whiteness historically involved much more conscious awareness of privilege and intentional actions to gain and maintain racial advantage, there is not the obliviousness to white privilege that is observed in the United States. Instead, other lines of thinking learned in the white family justify advantage and gain from an unequal and oppressive system (Steyn, 1999).

Frankenberg (2001) also pointed out that nowadays there may be a different kind of self-consciousness of themselves as a category for many white people, a sense that goes with what Frankenberg called "power evasive race cognizance" (p. 91). It involves a set of beliefs held by some white people that whites are disadvantaged by civil rights and affirmative action laws, that blacks have all the rights they need and would do just fine if they simply worked as hard as whites. The "power evasive" quality of such thinking is that the whites who think that way seem to be oblivious to the ways that they benefit from, support, and build a system that advantages whites over blacks.

Lest one think that white people who can talk or write about skin privilege and the obliviousness of others have escaped from obliviousness, Crapanzano (1986, p. 23), writing about whites in South Africa in the apartheid era, pointed out that feeling horrified and disgusted by an oppressive system from which one benefits may be a way to live with oneself while benefiting from the system.

Family Obliviousness to News of Impending Danger

Families often seem oblivious to news of impending danger, news of direct, material threat. Yes, most of what there is to know is irrelevant to most families, but what is interesting is that many families also seem not to know about impending events that could affect them very directly. For example, a family may be oblivious to news about global warming—what it is, how it is increasing, what it will do to the climate and the ocean levels, and the very substantial impact it may have on families living where the family lives. For a family anywhere, global warming research and theory offer very bad news (Engeler, 2006, November 3; Gelbspan, 2004; Goodell, 2007, November 1). Where I live, in Minnesota, they speak about a decline in rainfall and an increase in heat that will undermine agriculture, destroy forests, and bring new and dangerous insect infestations. They talk about more frequent and more destructive tornadoes and extremely heavy rain and lightning storms. They talk about life-threatening summer heat and the advent of diseases once limited to regions far to the south of us.

A family may also be oblivious to information about poisons, carcinogens, and other dangerous materials in their environment—such as the pesticides and hormones in their drinking water and food, the mercury in the air they breathe everywhere they go, the chemicals leaching into household air from materials used in building homes, the mold in their houses, or the proportion of drivers on the road who are impaired by alcohol, drugs, dementia, fatigue, or blindness. Why not know about these matters? It seems in the best interests of family members to protect their personal health and the health of their loved ones. Perhaps one reason families are oblivious to many dangers is that there are people who have considerable power who do not want them to know about these environmental dangers—for example, the makers of pesticides and pharmaceuticals, the operators of incinerators, the executives of electric power companies, mining companies, and major food companies, or the manufacturers of mobile homes.

Similarly, a U.S. family may have long delayed learning about the substantial devaluation of the dollar relative to the Euro and many other currencies. If they travel to Europe, Canada, or many other parts of the world or think about why petroleum products or imported fruits are so expensive you would think it would be in their economic interests to learn about dollar devaluation. If they did learn about dollar devaluation, they might be able to do things that would head off some of the most severe financial consequences of that devaluation for themselves. And they might start questioning politicians about the policies that have led to the dollar devaluation.

Families can find, if they choose, information on a number of impending threats to their well-being. Despite selectivity and censorship in the mainstream media, there is such a diversity of media outlets, so many alternative media outlets with so many differing viewpoints and interests, that people who want to learn much more about something than what is typically provided in the mainstream media can generally learn a great deal (Chomsky, 2000). But I believe many families do not look for such information, even on topics that would seem to be of great importance to them, do not find it, or do not pay much attention to it if they run across it. Perhaps the news of global warming, toxins in the water, or dollar devaluation is too threatening to some. The psychological pain, anxiety, and panic could be too difficult to handle, or the changes required to cope with the dangers would be hard for a family to engineer and might mean that new and daunting problems could be created for the family. News of impending danger may make a family's past and current commitments and ways of living seem a mistake, and that could be very unpleasant to confront. The news may seem to call for changed thinking and changed ways of living, and changes that substantial may be so difficult that families choose to be oblivious to the news. For example, at the height of the threat of nuclear war between the United States and the USSR, many U.S. families seemed to be oblivious to the magnitude of the threat that could have led to their obliteration (Wetzel & Winawer, 1986). Perhaps the obliviousness came from the sense that to face the threat would require greater changes in how they were living than most families were willing to make. On the other hand, the nuclear war example may be a poor one, because many projections of what nuclear war might do to the planet made it seem that there was no way to prepare for it. There was nowhere to hide. However, I would argue that obliviousness even to great threats has its roots in an investment in the status quo, not in evaluating possible changes in order to cope. So I think obliviousness to nuclear war danger came from people feeling locked into a value position that said the status quo for themselves and their family was best (where they lived and worked, what they did with their everyday life). This created an inertia that made major change impossible.

There are other ways of understanding obliviousness about threatening current events that seem deeply relevant to a family's well-being. One possibility is that some people with great power do not want ordinary families to know this or that and are successful at keeping people ignorant. So even with information in the news about global warming, mercury in the air, or the dollar devaluation, powerful forces are at work (for example, energy companies, political leaders, leaders in the investment and banking industry) that do not want people to decrease their current consumption of energy, to demand that energy companies make costly changes in how they produce energy, or to demand changes in the regulation of the financial industry. They do not want to change who has economic and political power in this country. They do not want enormous changes in how people invest or save their money or what they purchase. They do not want to end the globalization of the economy, the high rate of consumption of petroleum, or the high levels of consumer purchasing and indebtedness and the government borrowing that have fueled the devaluation of the dollar. These forces push families to be oblivious to many threats to family well-being. Then it is not only something like the family equivalent of ego defenses that underlies shared family obliviousness but also family response to the voices and actions of powerful institutions and people.

Another possibility is that family members share obliviousness to threats that they think they cannot do anything about; their focus is on what they can do. So, for example, they would pay attention to global warming if they thought they could do something to reduce it, head it off, or defend themselves from it. But there may be powerful forces at work to keep them oblivious to what they can do. Perhaps no family can have a measurable impact on world production of greenhouse gases, but some families can move to locations and housing where they would be safer from, say, hurricanes, tornadoes, and life-endangering heat waves.

A final possibility is that even information that could be very useful can challenge the family information priority system, and the priority system is resistant to change. Changing the family information priority system could create considerable uncertainty, disrupt what seems to be working, threaten basic assumptions about how to live, and threaten the normal hierarchy and patterning in the family. If so, we could say that families ignore news of global warming and other threats because it is difficult to change attentional priorities.

Family Obliviousness to History

Based on the discussions of shared family obliviousness so far in this chapter, it seems safe to assume that many families are oblivious to substantial parts of the history of the groups and political entities to which they belong. For example, many Euro-American families are likely to be oblivious to the history of white privilege. In that history would be rich information on cruelty and injustice in Euro-American relationships with American Indians; the history of slavery, the Jim Crow era, and contemporary expressions of racism against African Americans; the history of Mexicans living in what became part of the United States in the nineteenth century and the history of Mexicans who have come to the United States to work and perhaps settle;

the history of the Chinese in the United States; the internment of west coast Japanese Americans during World War II; and, in fact, the entire history of all groups that have been badly treated in the United States. They are also likely to be oblivious to how the history they learn and their ideas about family are linked to the history of oppressive actions by Euro-Americans (Chambers, 2001, pp. 36–41). White home owners may typically be oblivious to how their land came to be available for home building and ownership—how it might have been where American Indians lived, traveled, and drew sustenance. As this chapter has indicated, obliviousness to history could be based on many factors, including not caring and not wanting to feel guilty.

Obliviousness that is maintained collectively over many years writes history. If nobody paid attention to what the Spokane Indians said about the fur traders and missionaries who transformed their world in the first part of the nineteenth century and if those who wrote the earliest histories of the Spokane were oblivious to the Spokane realities, we reach the present with an irretrievably oblivious history of the Spokane. We will have a history, but one that is ignorant or possibly very wrong about a great deal of what the Spokane thought and said.

The United States is not necessarily different from other countries in historical obliviousness. For example, in Nicaragua, the Contra War, which led to many people "disappearing," has been generally treated with a great deal of silence at all levels of society (Tully, 1995). There are those in Nicaragua who resist the forces for obliviousness, for instance, some women whose relatives were "disappeared," and that may be an important point about counter-obliviousness forces in systems, that confronting silence does not necessarily end the silence. Silence can have an awesome power to continue at family and societal levels, no matter what information confronts it and what voices speak up. In its uncommunicative continuation it can be difficult to know whether silence represents obliviousness or something else (cf. Cohen, 2001, p. 9), perhaps a mute approach to all the world or a fear of speaking up about what one is fully aware of.

In terms of the dynamics of silence and obliviousness, the United States may not be so different from other countries where the government denies or hides much. Here there is, for example, government destruction of information (paper files shredded, electronic files erased, reports removed from the Library of Congress and other archives). Here there is the maintenance of secrecy with millions of government documents classified as secret each year, including, apparently, many that are secret only because revealing their contents would embarrass someone in government (Baker, 2002; Podesta, 2003; Rogal, 2003). Some of those documents may eventually become public knowledge (e.g., Farah, 1999; Kornbluh, 1999), but responsible (or perhaps I should say, "irresponsible") government officials would have been safe from public scrutiny for years. Here government scientists are muzzled so they do not reveal information that is embarrassing to government policy makers (*Seattle Post-Intelligencer*, 2007, March 8). Here government officials feign ignorance of poverty and injustice in order to maintain policies that benefit the wealthiest and most powerful (Limbert & Bullock, 2005). Here there are government officials who work hard to be oblivious to what happens to individuals and families as a result of legislation and legislative neglect. Here there is corporate destruction of embarrassing documents (Weissman, 2002), as well as corporate manipulation and control of the news media, so the news media say nothing or give only a distorted view about, say, a major corporation's violation of labor laws or selling of defective and dangerous products. Here mainstream media news accounts distort in ways that support the power and reputation of those currently in power (e.g., Goodman, 2007, writing about contemporary accounts of the Nixon and Ford administrations). Here there is the naming of events in ways that contradict responsible views of what has been going on (Herman, 2006; Kolhatkar & Ingalls, 2006), and these ways of naming not only manipulate thought and attention, they can rewrite history and morality. Herman (2006) focused on the names given to aggressions committed by the United States, its allies and client states, but one can focus on misleading, obliviousness-promoting naming in many other contexts. For example, a law with the benign name "The No Child Left Behind Act" includes legislative mandates that from some perspectives increase the chances that some, perhaps even all, children will be left behind from the perspective of some educational standards (Doster, 2007; Emery & Ohanian, 2004). Woodward (2006, November 27) pointed out that many words and concepts used by government officials in communicating with the news media are calculated to keep the public from knowing much. For example, "hunger" becomes "food insecurity," "torture" becomes "self-injurious behavior incidents," and the tax on a small number of multi-million and billion dollar estates becomes "the death tax." And the obliviousness that government promotes may leave many families in a more precarious situation than they might otherwise occupy. For example, Fox (2001) argued that widespread ignorance in the United States of federal tax policy means that many families are taxed unjustly, the government loses considerable revenue, and pension systems do not help many older people to live above the poverty level.

Another way in which news reports do not provide all the important news is that government or corporate whistle-blowers are at such great risk (Bennett, Lawrence, & Livingston, 2007; Martin, 2007) that many people may never try to tell others about government or corporate wrongdoing or corruption that they know about. Government and corporate "whistle-blowers" are discouraged from talking to reporters (Soley, 2002). Despite the existence of what seems to be legal protection for whistle-blowers, whistle-blowers or people suspected of whistle-blowing are at considerable risk of losing their jobs, often fired on trumped-up charges (e.g., Blumenthal, 2006), and quite possibly blacklisted so that they are unable to obtain decent jobs elsewhere (Soley, 2003). According to Sandler (2007), roughly 97 percent of U.S. government whistle-blowers who seek protection from federal whistle-blower laws are denied such protection. Whistle-blowers may be harassed, ostracized, slandered, referred to psychiatrists as though they are mentally ill, demoted, and otherwise ill-treated (Alford, 2001, 2007; Martin, 2007). Furthermore, by making the whistle-blower, rather than the allegations made by the whistle-blower, the subject of investigations and news reports, the matter to which the whistle-blower was calling attention may become a side issue or even ignored (cf. Bennett, Lawrence, & Livingston, 2007, pp. 156–61; Martin, 2007, pp. 67–68). Also, in the United States, news about many potential or actual scandals is suppressed because there is congressional refusal to create news by investigating various matters or investigating them deeply and thoroughly, for example, what underlies the gigantic profits of certain corporations. Nor is the public innocent in accepting the forms of historical obliviousness offered it (Cohen, 2001, p. 11). If the public does not challenge the secrecy, misleading labeling, control of the news, selective reporting, distortions of what happened in the past, lies, and suppression of whistle-blowing, it is colluding in creating and maintaining its own obliviousness.

Ignorance about Other Families

Each family lives in the context of other families. Families pick up information, ideas of how to behave, approval and disapproval, ideas of what normal family life is like, and frames of reference from other families. But families may often be ignorant of what goes on in other families. It is partly a matter of privacy and of each family keeping much information from others. In fact, for many in the United States the ideal relationship with neighbors may be a minimal one, particularly if the neighbors seem to differ from them in culture, language, or social class (Benson, 1990). They may not want to know much about neighboring families. That could be understood in many ways. People who are different can be unpredictably or predictably uncomfortable to be with. And then there is the fact that by maintaining obliviousness to neighbors a family is free to pay more attention to other matters, free to attend to information that for them is higher priority.

If neighbors do not want to know about neighbors, one can talk about pluralistic ignorance, an ignorance in which everyone, every family, every individual, is ignorant of the others, and in that ignorance families and individuals may make ignorant assumptions. Each family may, for example, assume that they are like other families, when they are not. Or they may assume that they are different from other families, when they are not. Believing they are different from other families, they may work at hiding what they think is their deviance from others. Willits (1986) reported that adolescents in her research grossly overestimated the conflict that other adolescents had with their parents, so there was pluralistic ignorance with most adolescents perceiving their family as having deviantly little parent-adolescent conflict. In fact, most families had little parent-adolescent conflict. Possibly the pluralistic ignorance had no impact on the families involved, but perhaps it made the relatively rare family with considerable parent-adolescent conflict feel normal and the many families in which there was little parent-adolescent conflict feel that they were doing better than most families. So the pluralistic ignorance about parent-adolescent conflict might have led to good feelings in most if not all families.

Conclusion

Family members share obliviousness to much of the family's context, to events nationally and internationally, to history that is deeply relevant to their current situation, to the trauma of others, and to their neighbors. Families who benefit from an oppressive system seem often to be oblivious to that system, how it harms others and how it benefits the family. Families are probably generally oblivious to their obliviousness, to their very substantial part in producing and maintaining their obliviousness, and to the powerful forces in society that intend them to be oblivious about many matters. One might think that all this family obliviousness saves attentional resources so that a family can be aware of much if not all that goes on in the family. But as the next chapter argues, family members generally share obliviousness to much that goes on in their own family.

4

Obliviousness to Matters
within the Family

Just as family members share obliviousness to most that goes on in the world outside the family (see chapter 3), they also share obliviousness to most that goes on in their own family. There is vastly more going on than they could possibly know. So, for example, as a family they may not usually attend to what family member X does when alone in the kitchen or bathroom, may not attend to what family member Y is reading, and may not inquire much of the time about what anyone in the family is thinking.

In everyday life, families may ordinarily be oblivious to their moral inconsistencies. For example, they may say that they value generosity and helpfulness with regard to others and act on those values in terms of charitable donations, but they may also live at home in a way in which they are anything but generous and helpful to each other with regard to, say, who does the kitchen chores and housecleaning, and they may be oblivious to this inconsistency. Or, to cite another example, they may take a religious stance that values charity and compassion for the poor, and yet they may also, in their daily life, ignore the poor, and to this too they may be oblivious.

From the perspective of an outside observer of families, family interactions are ordinarily predictably patterned (Rosenblatt, 1994). Family members may be aware of and quite articulate about some of their patterns, for example, an adult couple's going-to-bed routines (Rosenblatt, 2006). But, as the interpersonal communication, parenting, and family intervention literatures suggest (e.g., Burgoon, Berger, & Waldron, 2000; Dumas, 2005; Horton-Deutsch & Horton, 2003; Long, Angera, & Hakoyama, 2006), individuals in couples and families can apparently be quite oblivious to crucial aspects of the patterning of their interactions. Sometimes they do not seem to be aware at all of a recurrent sequence of family interaction, for example, their pattern of getting into conflict or of silencing one another. Although these interpersonal communication, parenting, and family intervention literatures generally focus on individual processes, especially individual mindfulness/mindlessness, it seems plausible that mindfulness/mindlessness to family interaction pattern would often be shared throughout a family.

Perhaps in the United States another common area of shared family obliviousness is for each member of a family to think that his/her family is much more a group of freely functioning individuals and much less interdependent and closely connected than they are. In individualist cultures of the United States it is possible for the family as a whole to be oblivious to the many ways that the self that each member thinks she or he has is only what it is because of links and interactions with other family members. Among the reasons why such shared family obliviousness, to the extent that it exists, could be a problem is that the family connections to which each family member is oblivious may be destructive, hurtful, limiting, and in other ways less than ideal (Hankiss, 2006, pp. 129–30).

The lessons of shared family obliviousness to matters within the family begin in early childhood. It is at home that children first learn family obliviousness rules; as their curiosity is shaped, their attention is directed to some matters and not others, and some of their questions are not answered or are answered tersely, evasively, incompletely, or not at all. Doors are at times closed to children, and they may realize that some topics are intentionally talked about out of their hearing. As children grow up they also have the opportunity to see ways in which their family elders are oblivious to matters outside and within the family, and so they see the adults modeling obliviousness. Assuming that family socialization for obliviousness to many matters within the family is effective, by the time a person is an adult, the person can be counted on to miss a lot of what goes on in his or her family.

To further make the case for the idea that families and family members seek obliviousness, one can cite instances where, if family members are not oblivious about some matters in the family, they long to become oblivious. Consider, for example, parents of little children. In many families parents feel that little children must be watched closely lest they harm themselves, others, or family property. Many parents who watch a young child closely long for the day when the child can be trusted to be left alone for a few minutes, freeing the parents to be oblivious to some of what the child does. And parents often find ways of gaining that obliviousness temporarily while the child is still not to be trusted—for example, putting the child in a playpen, highchair, or swing, hiring a baby sitter, or getting the child to take a nap.

Necessity and Costs of Everyday Family Obliviousness to Matters within the Family

Everyday family obliviousness to matters within the family probably works for a family most of the time by protecting the family from information overload, by keeping the family working on what is highest priority, and by not creating any serious problems. But once in a while the obliviousness means that things are ignored that might better be attended to for the well-being of the family or of individual family members. One example would be a family member who develops symptoms of a serious illness that everyone ignores until the condition is life-threatening. Let's say that someone in the family has had a stroke. The person's speech is suddenly slurred; the person's hand is suddenly weaker; the person's memory has suddenly deteriorated. But the person and other family members miss the symptoms, or if they notice the symptoms, they may think of them as signs of fatigue or ordinary aging. Recognizing immediately that the person may have had a stroke could lead to earlier and more effective medical treatment and to a higher quality of life for the individual and the family. Recognizing that the family member has had a stroke could lead to better family help, understanding, and support for the individual. Other examples of ignored information that by some standards might have been better to pay attention to would include family obliviousness to the family member who has become addicted to something (more on that later in this chapter), the family member who is stealing from the family, the family member who has become extremely depressed, the family member experiencing severe emotional or physical pain, or the family member who has developed anorexia. With any of these problems, the family may pay a cost that varies with the problem but that can be quite heavy and long-term. Similarly, obliviousness to moral inconsistencies may sink a family into a hypocrisy that corrodes and cheapens all its moral values, and that too may have dire consequences in terms of life meaning and moral comfort. Still, shared family obliviousness is probably often not a problem. It may be helpful not only in defending against information overload but in family coping with difficult problems. For example, obliviousness can help family members to cope with a family member's head injury, when obliviousness can lead the family members to constructive activity and to supporting hope regarding the head injury (Ridley, 1989).

Patterns of Conflict, Making Up, and Conflict Avoidance

Family systems theory alerts us to the pervasiveness of patterned interactions in families (Rosenblatt, 1994). In some couples and families there are recurrent patterns of conflict, and perhaps verbal or physical abuse, and then making up. One couple I interviewed (Rosenblatt, 2006, p. 100) said that they had been oblivious at one time that they had a pattern of extremely harsh arguments on nights when they had both been drinking alcohol. Eventually, they came to realize that there was a pattern and agreed that when they entered into one of those alcohol-fueled arguments they would stop, disengage, and sleep on it. In the morning, when they were sober, the issues, the anger, and the conflict had typically evaporated.

Couples and families may enter into recurrent, patterned arguments in many circumstances—for example, when one of the partners is tired, at certain times in a woman's menstrual cycle, at certain times in the pay cycle, on vacations, when a school-aged child is home sick, while traveling, on certain holidays, on their wedding anniversary, following sexual intercourse, on weekends. They can know fully that they are in conflict or have just ended it and may remember that they have had similar conflict on similar occasions, but they still may not catch on to the patterning of it (e.g., Derdyn & Waters, 1981; Dumas, 2005; Feldman, 1980).

Making up in order to end conflict is also patterned in some couples and families (Rosenblatt, 2006, pp. 101–104). And, as with conflict itself, the participants may be unaware that there is patterning to when and how making up is carried out. For example, a heterosexual couple may be unaware that their pattern of making up is that typically on the morning after intense conflict the woman partner initiates peace by apologizing for something (whether or not, from an outside observer's perspective or her own she did anything that merits an apology).

Conflict avoidance may also be a matter of obliviousness, and typically the obliviousness is at two levels. Families may be oblivious that they are avoiding conflict, and then to maintain that obliviousness they may have to be oblivious to the issues on which they are avoiding conflict. Shamai and Lev (1999), for example, wrote about the relationships of Israeli couples living in the Golan Heights on the border with Syria. In the face of possible forced relocation by the Israeli government, some couples had frequent arguments and planned for possible forced relocation, but, interesting for the purpose of the discussion here, some ignored the possibility of forced relocation. In their seeming obliviousness to the possibility of forced relocation they also avoided conflict about the issue.

Obliviousness to Teen Activities and Sexual Orientation

In much of this chapter, the emphasis is on shared family obliviousness to matters within the family. However, there are interesting and important situations where obliviousness is not shared family-wide. One of them is in the relationships of teenagers with their parents. In many families in the United

States an important part of growing up is to become increasingly independent of one's family in the sense of more often thinking independently and doing things outside of parental purview (e.g., Bloom, 1980). For that to happen, parents ordinarily have to become increasingly oblivious to aspects of a ch.ld's life, perhaps innocuous matters such as playground conversations or what is written in classroom assignments, but perhaps less innocuous matters by the standards of some parents, such as sexual and drug explorations. At the same time, part of growing up for many children is to learn how to keep their parents oblivious to some of what they do, feel, and think. That is, growing up involves increasing the areas that one keeps private from parents and getting better at keeping one's parents oblivious. So a teenager may feel more grown up, and by some standards be more grown up, if she or he does things that parents do not know, if parents do not express curiosity about so much in the teenager's life, and if the teen is at times successfully deceptive.

Small wonder, then, that when I Google phrases such as "oblivious family" and "family obliviousness," some of the hits are to someone writing about how her or his family was oblivious to certain things that the writer did as a teenager. The teenager might have been sexually active, using drugs, bulimic, cutting, skipping school, smoking, or engaged in other activities that presumably the teenager's parents (1) would not want to go on and (2) should ideally know if, by their own standards, they were properly attentive to the well-being of their child. There also are a substantial number of hits to accounts of parents being unaware that a teenager had a lesbian or gay sexual orientation. In these writings there seems to me to be a sense that the information was present for the parents to know if the parents cared to.

In many of the Googled cases where a writer talks about family obliviousness during her or his teen years, the writings seem to me to be critical and sometimes even contemptuous of parental obliviousness. There were implied or stated questions such as, "If these people cared about me, how could they be so oblivious?" "If my parents even observed me casually, how could they miss that I was cutting and sexually active?" And yet, the teenagers were, I presume, trying to keep parents in the dark. And the parents may well have been trying to help their teenagers keep them in the dark. The parents, as good parents trying to help a teenager to become more autonomous, were colluding in becoming oblivious to areas of their teenager's life. The teenagers may have been allowed more privacy and more time away from the parents—perhaps having bedrooms to themselves with doors that could be closed and even locked, being allowed to be in the bathroom with a closed door, allowed more time away at school and with friends, perhaps given a private telephone, perhaps given use of an automobile, and given little of the monitoring that a young child would receive. So parent obliviousness was partly a matter of respecting the teenager's movement toward adulthood and feeling little concern that the teenager would get into the kinds of trouble that a small child might. The obliviousness was also partly a matter of the teenager working at achieving distance from parents—establishing increased independence and wanting greater freedom from parent monitoring and control. So despite the criticism and contempt in some Web entries, parental obliviousness might well have been desired by the teen. But as the criticism and contempt imply, some teens might not always have expected to be successful or wanted to succeed so thoroughly at keeping parents oblivious.

All this suggests an interesting perspective regarding closeting from and coming out to family. For some people, coming out to family is not only about revealing sexual identity; it is also a commentary about obliviousness in the family. The person coming out may not only have whatever emotions might be focused on creating and arriving at a new personal identity in the family's eyes and risking family sorrow, anger, and retaliation. The person coming out may also have feelings about how oblivious the family was and how her or his attempts to keep things private and the family's respect for that privacy had in some ways been disappointing, although also welcomed. Then, to parallel the young person's reactions, the family's reactions to the coming out might in part reflect their feelings about having voluntarily been oblivious and having been allowed to be oblivious. Perhaps the same set of complexities might be present when a teenager or young adult reveals a history of other activities that might upset parents. But perhaps the intensity of reactions is often greater in the case of sexual orientation, partly because sexual orientation is such a sensitive issue for many in the larger society and partly because it is central to self for many people.

Family Obliviousness to Alcoholism or Drug Addiction

There are families in which a member recurrently is under the influence of alcohol or drugs and seems to be addicted and where, from an outsider perspective, nobody in the family seems to know the extent of the usage (Hermida, Villa, Seco, & Perez, 2003). Alternatively, although some family members eventually become aware of the pattern, at first nobody seems to notice the pattern, considers the pattern significant, or considers that what is going on is an addiction and is harmful to the individual and the family (Jackson, 1962; Krestan & Bepko, 1993; Metzger, 1988). I sometimes encounter members of families in my research interviews who have realized, often along with other family members, that someone in the family has a problem habit. Some individuals say that they had long been oblivious to the pattern, as had the other members of their family and perhaps especially the person with the problem habit. People say things such as: "We thought it was normal"; "He drank in private, so we didn't have any idea how much he was drinking"; "We only knew he had a problem when he was fired from his job"; "We only knew she had a problem when she entered the treatment program"; "I didn't think anything of it until I started therapy"; "He still doesn't think he has a problem."

Problem habits may be noticed by family members, but the meanings of the habits may be a matter of collective family obliviousness (Krestan & Bepko, 1993). Why? Possibly the person with the habit does not want to be stopped or challenged, and especially if the person is effective at concealing some or all signs of the habit or is someone with the power to make things difficult for others, family members might be inclined to believe that there is nothing to challenge. Then too if the habit has long been present, it is difficult to notice that there is something there to notice. It is just the way things always are.

At another level, family obliviousness might be linked to the possibility that challenging the problem habit could rock family relationships. Everyone in the family has their place, their sphere of operation, their escape hatches, their familiar ways of behaving, thinking, and feeling in relationship to the problem habit. Challenging the problem habit may threaten all of this. So even if one person starts to bring the problem habit to light and to alert and inform other family members, everyone else may resist awareness or further information, because they have much to lose. They may even fear that challenging the problem or breaking the silence could shatter the family (Krestan & Bepko, 1993).

Another mechanism of maintaining a problem habit might be called replicative secrecy. The person with the problem habit is secretive in ways that push other family members to be secretive, and each person's secrecy pushes the others toward even more secrecy (Krestan & Bepko, 1993). Thus, what started out as denial by one person may become a family-wide pattern of denial reinforced by secrecy and unwillingness to challenge. The processes underlying the denial may be different for different people. The family member with a problem habit may deny it out of shame or out of anxiety about what would happen if things would change. Others in the family may deny it out of fear of the wrath of the person with the problem habit or fear that family relationships will become more difficult if the problem habit becomes a topic of contention. So together the family members, possibly with varying motivations, work out ways to be oblivious to how much there is a pattern, how much harm is being done, or even that there is anything going on that merits attention.

The picture that emerges is not of a happy or satisfying family, nor one with a great deal of emotional sharing or real intimacy (Krestan & Bepko, 1993). In Krestan and Bepko's view, it is ironic that along with the great fear that the family will be shattered if the problem habit is challenged or even discussed there can be, with so much family obliviousness (my word, not theirs) and reluctance to speak up about a problem habit, a deterioration toward being a family that does not provide family members with very much emotional support or nurturance.

From a social construction of reality perspective, people work out much of their sense of reality in interaction with others (Berger & Kellner, 1964; Berger & Luckmann, 1966), and those others are often close family members. It is difficult to make sense of events or feel confident about the sense one makes of events if one comes to that sense without interacting with others. According to this line of thinking, one result of family obliviousness about the alcohol or drug problems of a family member is that if one family member becomes aware of the problem habit but has nobody else in his or her oblivious family to talk with about the habit, the person will have trouble coming to a sense of what is real about the problem. Isolated from interaction with others about the problem, the person will have trouble deciding whether there is a problem, what to do about the problem, what the consequences are of the alcohol or drug usage for self and family, and even whether the alcohol or drug usage is bad or good. Thus, there can be a heavy cost to becoming nonoblivious to a problem habit in an otherwise oblivious family.

Incest and Sexual Abuse

Incest and sexual abuse may occur in secrecy, but often there is some evidence present for others in the family who are not directly involved in the incest or sexual abuse to know if they choose (e.g., Hitchens, 1972; Hoke, Sykes, & Winn, 1989). It may be evidence that somebody is hiding something; it may be certain sounds; it may be a child's depression; it may be a pattern of inappropriate touching and words, it may a child's bleeding from vagina or anus. But I gather that in some families in which incest occurs everyone but the perpetrator and the victim is oblivious. And even one or both of them may enter a state of denial that borders on obliviousness (Trepper & Barrett, 1989, p. 108).

Obliviousness to incest or sexual abuse is no doubt driven in some families by fear (cf. Wright, 1991). A wife may fear losing her husband and her marriage. Anyone may fear family conflict (Hoke, Sykes, & Winn, 1989), family disruption (Hoke, Sykes, & Winn, 1989), family change (Trepper & Barrett, 1989, p. 108), financial loss (cf. Wright, 1991), what others outside the household might think, or involvement with law enforcement and child protection authorities. Obliviousness may also arise out of concern that the family lacks capacity to assimilate and react appropriately to knowledge of incest or sexual abuse. Giving up one's obliviousness to incest or sexual abuse may threaten one's place in the family, everyone else's place, and perhaps the continuing existence of the family itself. If Husband/Dad is the perpetrator, will someone or everyone knowing and speaking about what he has done mean that he will have to leave the home? Will he have to go to jail, with all the shame, relationship disruptions, and financial loss that might entail?

From another angle, incest and sexual abuse are facilitated by abusers having resources, such as housing with enough privacy to keep others oblivious (Fraad, 1996/1997). U.S. housing, particularly for the well-off, provides a shield of physical privacy to keep others oblivious to incest and sexual abuse. Furthermore, U.S. families are defined in ways that create a shield of entitlement that makes children into parental property (Fraad, 1996/1997) and as a consequence keeps others relatively uncurious about what a parent does with a child. More generally, the idealization of the nuclear family in the United States and its ideological separation from the outside world makes it easier to keep outsiders, and in some sense insiders, oblivious to terrible things that happen to children (Fraad, 1996/1997). In fact, the idealization of the family makes it difficult for the nation to be aware of the incidence of incest and sexual abuse, their terrible cost, and the desirability of legal and other changes that could help to protect children from incest and sexual abuse (Fraad, 1996/1997).

Obliviousness to the Crimes of Family Members

There are families in which someone has committed a crime and others are oblivious—perhaps crime families or the families of drug dealers, families of corrupt business executives, families of pornographers, or families of tax cheats. It is interesting to consider where the obliviousness comes from and what uses it serves in the family. Perhaps the strongest literature in this area deals with families of Germans who murdered millions during the 1930s and 1940s.

Hecker (1993) wrote about the obliviousness of children and grandchildren in post–World War II German families to the bloody activities of their elders. In fact, there is evidence that even during the war, even in the families of those who had a central role in murdering millions, the family life of murderers seemed banal, with family members seemingly oblivious to the murdering (Katz, 1993, pp. 72–75). But obliviousness can have its costs. Following World War II, for German families who seemed oblivious to what family members did during the war there could come a great deal of family difficulty (Hecker, 1993)—avoidance of conversation topics, nonverbal communication not to talk about certain matters, messages from the parents and grandparents that their only hope was in their children and grandchildren (for redemption? to erase the past?). There were also messages to the children about the dangers of political involvement, and for people of every generation in the family there were issues of who can be trusted. All this is linked to obliviousness in the sense that the avoidance by older family members of certain conversation topics might well lead younger family members to be oblivious to what their parents and grandparents did. Avoidance of topics related to the war might have led younger family members of Holocaust perpetrators to work at not knowing about the Holocaust and World War II in any detail, that is, to work at being oblivious. And in their work at being oblivious they may have been as active as members of the older generation at avoiding certain topics and at maintaining a family pattern of silence about matter related to wartime activities (Rosenthal, 1998; Rosenthal & Volter, 1998).

One might assume that such obliviousness would facilitate intergenerational intimacy, that younger family members oblivious to their parents' and grandparents' roles in a monstrous history would find it easier to be close to their parents and grandparents. But that did not always happen. The secrets, evasions, myths, and denials that helped to maintain the obliviousness of younger family members could also produce considerable emotional distance between the generations (Krondorfer, 1994; Stierlin, 1981, 1993). Quite possibly, even if one wanted to bridge the emotional gap, the history of distance, evasion, silence, and concealment would leave a person uncomfortable about trying to begin to bridge the gap, because it would not be clear what to ask or even what to say (Schindler, 2005). Alternatively, the obliviousness might be accompanied by enough silencing (Zuk, 1965) and denial that it would be evident to family members that there were family strains (Zerubavel, 2006) even as they remained oblivious to the topics about which there was silencing. Perhaps in some families there was not only obliviousness but resentment of silencing, evasion, and denial.

If efforts to maintain obliviousness produced so much difficulty for some families of former Nazis, why the obliviousness? As is argued in chapter 2, family members with power may use their power to control the knowledge other family members have. Among the many elements of such knowledge control is, no doubt, the power to keep others oblivious to the sins of those with power. With power, one has tools to maintain one's power and reputation. So possibly the obliviousness and all that it cost some German families was linked to family members in power in the family trying to maintain a more positive reputation in the family than they would have if what they had done was known throughout the family. And perhaps it is the same with crime families, drug dealer families, and other families of wrongdoers, that reputations in the family are maintained by keeping others in the family oblivious. That might not be the only reason for obliviousness in any family with a wrongdoer, but it may be a key underlying factor in many.

Denial of Death in Families

People are said to be inclined to deny death (Becker, 1973; Liechty, 2002). According to Becker (1973), denial of death involves struggles of the individual, quite possibly unconscious ones, to deal with personal mortality. It is the individual who does not write a will or who takes great chances, as though immortal. It is the individual who does things to try to leave a legacy of some sort (for example, children, works of art, literary works, a garden, successes, or fame) as a kind of immortality. It is the individual who may not mourn a death or aspects of a death, because to do so is to be confronted with the intolerability of mortality.

Although much of the literature on denial of death focuses on the individual (and in some ways on the entire society or species), I think a case can be made that denial of death often involves shared family obliviousness. Individual denial of death can be much more easily sustained if everyone in a family supports the denial, quite possibly by denying approaching death, since they have their own death anxiety to deal with plus their anxiety about the death of others in the family. Furthermore, part of dealing with death anxiety is to seek "social support and the safety of fitting into the social group" (Liechty, 2002, p. xi). So with denial of death one would expect that often the whole family functions as though oblivious to personal mortality and the mortality of one another. Included in that obliviousness might be a pattern of not talking about matters related to death or minimizing such matters if they come up.

However, there are ways of dealing with impending death that look like obliviousness but arguably are not. Sometimes a person who is dying is aware of approaching death but wants to continue his or her ordinary life without reference to personal death, seemingly oblivious to the impending death (Zimmerman, 2004). Then, I think, the person may not be denying so much as choosing to embrace what can still be embraced of life. Or perhaps the person is trying to minimize personal discomfort and the pain of others by not talking about what will be painful (Zhang & Siminoff, 2003). But then others close to the person who is dying must shoulder the burden of awareness of the impending death and have the responsibility of supporting the dying person's efforts to be seemingly or actually oblivious (Seale, 1995). Obliviousness or pseudo-obliviousness when someone is terminally ill will make it difficult for the dying person and family members to verbalize their feelings about the impending death and its possible aftermath for the survivors (Doherty, 1976). Perhaps some or all family members feel less anxiety by not talking about the impending death. Perhaps they can put off dealing with painful emotions and with challenging demands for new ways of thinking and acting.

Quite possibly, what families do when a family member is dying, whether to deal with it directly, to be oblivious, or to engage in pseudo-obliviousness, is a continuation of long-standing family patterns of dealing with death, illness, and the anxiety that might arise if any long-standing pattern were disrupted (Miller, Bernstein, & Sharkey, 1973). But there are costs to those long-standing family patterns. The costs of family obliviousness or pseudo-obliviousness to a pending death include lost opportunities to say goodbye, to talk about what the deceased would like to have happen after the death, and to deal in what might be a mutually supportive way with painful feelings concerning the pending death. The costs include all the energy that must go into efforts not to know or speak about the impending death. And the costs include individuals being on their own in constructing realities about death in general or the particular person's dying and death. Without mutual work at social construction about dying and death, without human-to-human contact about the death, people are likely to be less sure of what is real and true, less confident in their beliefs, and more at sea about how to deal with what is happening and is likely to happen. The obliviousness or pseudo-obliviousness could make family members more confused and anxious about death or the particular dying or death. It may make it harder for anyone to hold on to meanings concerning death. And it could mean that family members are not in a position to coordinate with each other—for example, in helping a dying family member to be comfortable or in having shared ideas of what to decide about keeping a family member on life supports if there seems to be little or no likelihood of the person being able to think or be aware again.

Looking at larger systems linked to a family, it could be argued that obliviousness to death can go on in some parts of the larger systems only because it is supported by what goes on in other parts of the larger systems. One imagines newspapers not reporting news that would threaten the obliviousness of politicians or the general public about death (for example, deaths caused by U.S. government action or inaction). One imagines governments keeping secret from the public what would threaten the public's sense of well-being in the face of the next possible epidemic or killer storm, rising seas, and food shortages that seem likely to come with global warming. Thus, obliviousness concerning death might often involve a conspiracy of many different people and institutions. Obliviousness about death or a specific impending death is not necessarily simply a matter of the psychology of one individual or one family but may well be about the functioning of a number of systems and institutions.

Denial of Loss in Families

Some families seem to deny a death or other major loss at the affective level (Derdyn & Waters, 1981; Paul & Grosser, 1991). In their relationships with each other they do not seem to grieve a loss that one would assume would have had a profound emotional impact. Their denial of emotional impact may go on for a long time, even for decades, and may be related to family rigidity and to serious mental health problems of at least one family member (Paul & Grosser, 1991). The pattern also includes denial or otherwise seeming to "ward off" other losses and disappointments, and that may lead to restrictions on family members' independence or separateness (Paul & Grosser, 1991). This form of obliviousness is, then, not to the fact of a loss but to the emotional meanings of the loss. The obliviousness may be difficult for a therapist to detect. An emotional blandness in the present may seem not to represent denial of a loss but that the loss at some time in the past had been appropriately grieved. A therapist may, however, discover the long-term obliviousness to the emotional meaning of the loss through family member accounts of how they have not grieved over the years and, through inference, from family rigidities, family limiting of independence and separation, family elusiveness in talking with the therapist about the loss, and evidence that family members have not interacted about the emotional meaning of the loss. If the therapist is correct about the denial of affective meaning of the loss, a therapeutic technique (for example, Paul and Grosser's, 1991, operational mourning approach) that moves the family toward dealing with the affective importance of the loss may be helpful. On the other hand, it can be a daunting epistemological challenge to decide whether denial of loss has occurred and whether a family is oblivious to the emotional meanings of a loss. The family members may not be able to get in touch with where they have been about the loss at other times and places. And family members may come from a culture in which the emotional processing of a loss is very different from that of the therapist's culture, so what seems to the therapist like obliviousness may seem to be full awareness to someone who knows the relevant cultural code (Rosenblatt, 2001, 2003).

It is also possible that what seems like denial of the emotional impact of a loss is an artifact of not talking about the loss or of trying to keep others oblivious to the loss or its meanings. If the loss is too uncomfortable to talk about, if it brings up thoughts and experiences that a family member would rather keep others in the family oblivious to, if knowledge of the loss or its meanings might threaten the safety of other family members, a family member might try to keep the information from others. That the information is about something emotionally potent and that keeping others in the family from the information also keeps them from emotions of great intensity could make for emotional distance in the family. From that perspective, the emotional distance is only secondarily about emotional cutoff. It is primarily about information cutoff. For example, the grandchildren of victims of Stalin's purges in the Soviet Union might have experienced an emotional cutoff from their parents and grandparents. But that might only be an artifact of the elders not wanting the grandchildren to know about experiences that not only were horrible but could be dangerous for the grandchildren to be linked to (Baker & Gippenreiter, 1998).

From a rather different angle, the rituals and ceremonies dealing with deaths can be understood to shape realities and give meaning to the death in ways that focus awareness some places and not others. This shaping can be reflected in how the narratives of bereaved people can be rich in addressing some matters related to a death and its aftermath and not deal at all with other matters (Rosenblatt, 2000, pp. 2–4). The shaping can be reflected in the cultural themes and motifs that are so often part of eulogies, obituaries, prayers at a death, and ways of symbolizing the death. They deal with what they deal with, and leave out a great deal. So a family's dealing properly, by their culture's standards, with a death may well require family obliviousness as well as family awareness.

From still another angle, some losses are kept secret from most or all others, so almost everyone is oblivious to them. For example, some people in the United States may not tell most others about a miscarriage or stillbirth (Rosenblatt & Burns, 1986). This means that they may not receive nearly as much emotional support as they could, and it also means that each woman, each couple, and each family who has a miscarriage or stillbirth may be oblivious to how many others have had such losses and been affected by them. One consequence of that obliviousness is that it may be harder to talk about or even acknowledge feelings about a miscarriage or stillbirth or to make sense of and deal with those feelings.

Obliviousness as Intergenerational Debt

Deaths can bring intergenerational debts. For example, after a parent dies an offspring may feel that she or he owes it to the parent to be a better person by certain standards, to maintain certain identities and practices, or to become more expert on family history (Barner & Rosenblatt, 2008). Among the things that may be owed to a family member who has died is obliviousness about certain information. This may be an individual matter, but it can also be a family system matter, with all members of the family owing it to, say, Grandmother that they remain oblivious to aspects of the family past, the existence of a brother of hers who did something evil, a love affair she had, or the like. In fact, the intergenerational obligation to be obliviousness may continue Grandmother's own intergenerational obligation to be oblivious, perhaps to how her parents became wealthy or poor, their experience of persecution, or an ancestor's shameful act. Obliviousness involving intergenerational debt may have aspects of denial in that something was known before people became apparently oblivious, even if all that was known was that there was something one should not ask about or be curious about.

An interesting aspect of obliviousness based on intergenerational debt is that it may be particularly resistant to being overcome. Not only do individual family members have their personal sources of resistance, and not only does the family system have its mechanisms for maintaining obliviousness, but the debt owed intergenerationally may be a powerful force in itself for maintaining obliviousness. To abandon what the debt seems to call for might seem to be an unforgivable break with the past. If one cannot negotiate with the dead, or if one feels that it is impossible to pay back the dead well enough for their sacrifices, or if the obligations to the dead arise from interactions in childhood that one's adult persona cannot touch, it would be very daunting to try to end actions based on perceived debts owed the deceased.

Obliviousness to What Is Good in the Family

Many of the examples in this chapter are of family obliviousness to what could be considered to be negative, alcoholism or crimes, for example, so obliviousness means that what is harmful, shameful, embarrassing, challenging, and so on is neglected. But obliviousness may also be to what is good in the family, to love in the family, or goodhearted helpfulness in the family, or to the ways that family members provide emotional support to one another. One sees this kind of shared obliviousness in couples who are in intense conflict or who are thinking of divorcing. They may so concentrate on what is wrong that, even though they disagree about much, they may share obliviousness to what is good in their relationship.

Obliviousness to the positive can have its costs. Family members who are oblivious to what is good may not appreciate one another or not realize that what they have is extraordinary and should be maintained and nurtured. One can imagine that obliviousness to the positive may come out of something like paying attention to what is wrong rather than to what is right, but it may also come out of obliviousness to contextual information. That is, a family may not know enough about other families to see that what goes on in their household is better than many families have and best not be taken for granted.

Mutual Obliviousness in Couples and Families

One of the interesting things to consider about family obliviousness is the phenomenon of mutual obliviousness, when family member A works at being oblivious to much about family member B and vice versa. People who ostensibly should be closest to each other and know the most about each other can be, because of their mutual obliviousness, strangers to one another in significant ways. Some of that obliviousness and distance might by some standards be desirable. For example, in couples there is often a sense that having some time apart and some privacy from one another is a good thing. Conversely, knowing an enormous amount about one another can be intrusive and controlling.

Mutual obliviousness is perhaps partly about how much information can be processed. That is, mutual obliviousness in a family might be partly about attentional economics, that the family system and each individual in it will function less well if members know too much about one another. One way this makes sense is to consider a married couple who has a new child. Such couples may ordinarily experience that they spend less time talking about each other and more time talking about the child. So the couple moves to greater mutual obliviousness in order to process the large amounts of information needed to keep the child healthy and cared for. And if the child has some special difficulty that requires even more attention, the couple would move to even greater mutual obliviousness.

As an alternative to an attentional economics view of family obliviousness, it is possible that a family system may function to keep family members oblivious to things that if known could break up the family or make for extreme difficulty in family relations. If family member X knew more than she does about family member Y's thinking, friendships, hygiene habits, religious views, voting, etc., family member X might find it intolerable to continue living with family member Y or feel so enraged and upset that family life, for everyone, might become very difficult.

People sometimes joke about or deplore that many couples seem more interested in each other early in their relationship than they are later on. They seem to move to greater obliviousness, or to an indifference that seems to promote obliviousness. One way to understand that is that in the earlier phases of their relationship each is gathering information in order to make a decision about whether to continue the relationship or each needs more information in order to figure out how to get along with and live with the other. But once the decision is made to stay in the relationship and once a reasonable pattern of getting along is worked out, they need not gather nearly as much information. Another mechanism that no doubt operates is that one needs little information about what one can rely on. If one's partner is reliable in the ways that matter to one, one can tune out much about the partner. New information might be nice, but it is not necessary, and so it might make more sense to invest information processing energies elsewhere (or nowhere, because of the risk of overload). In that sense, obliviousness in the relationship of a couple who has been together for a long time might well be a sign that things are going well for them. They may not need to process much information because they know what they need to know to function together.

On the other hand, mutual obliviousness in a family can mean that matters of very great importance are being missed and perhaps someone is being harmed as a consequence. Also, a couple or family relationship with a great deal of mutual obliviousness may be difficult and unrewarding or even punishing because there is not enough human contact and caring. One hears of couples in which one of the partners leaves because there is not enough emotional contact or communication. Presumably that could be understood as a negative effect of obliviousness, though it is possible that couples that break up in such situations would have broken up sooner without the obliviousness they had to certain matters. Also, it is not that the couples who have high levels of contact and mutual information lack obliviousness. They may have wonderful conversations about where they stand about, say, politics, love making, and flowers, but there is still much obliviousness about one another.

Family obliviousness rules and patterns may call for asymmetry in obliviousness, with, say, one adult partner much more oblivious to what goes on in the relationship or the household than another, or children oblivious to more that goes on in the family than parents are. And whatever goes on in family relations can be understood as in part about maintaining obliviousness. The spats, the ways they eat meals, their bedtime routines, their separate hobbies and recreations, their use of space in the house, interactions when they drive places together, and so on may give them information about much but will also keep them from various kinds of information about one another. This could be taken to imply that family members can feel close and can even feel that they are closer to one another than to anyone else in the world, and yet obliviousness operates to keep them informationally apart about many aspects of what goes on with each of them.

5

Shared Obliviousness and Family Decisions

In everyday conversations about decision making, the experts might typically say that the best decision is a well-informed one. But if being well informed means knowing a great deal about many options, is being well informed necessarily best? Might the process of becoming well informed combined with the process of deciding what to make of the information, combined with the uncertainties and subjectivities inherent in many decisions slow down, or even block decision making (Schneider, 2006)? Might obliviousness, even intentional obliviousness to many decision-relevant matters, facilitate decision making? It is possible that obliviousness is often a valued and useful part of decision making (Schneider, 2006).

Whether or not obliviousness facilitates making the best family decisions, it may well facilitate the process of family decision making in terms of how much time is put into research, information evaluation, family discussion, and family decision. People ignorant of many decision alternatives and ignorant of many of the potential consequences of the alternatives they are aware of probably find it easier to make decisions.

From another angle, on many decision matters, the family decision process may protect areas of shared family obliviousness. For example, a family deciding about where to go for a winter vacation may want to maintain its shared obliviousness to poverty and oppression in Latin America or the Caribbean and so they might want only to consider winter resorts in these areas that are isolated paradises. Thus, a desire to maintain shared family obliviousness might shape and limit certain decisions.

Not only might obliviousness facilitate decision making, it might also shape the evaluation of the quality of decisions. Over the months and years after making an important decision, family members may reflect on that decision. And quite possibly they will not want to know the ways the decision could be seen as a mistake. For example, they may not want to know how buying their SUV has harmed the earth. Thus, family members may work together to be oblivious to the ways that their major decisions may not have been good by some criteria.

Obliviousness and Decision-Making Quality

When a family enters the process of buying a house, deciding to have or adopt a child, or deciding to save for retirement, they are almost certainly oblivious to a great deal of potentially relevant information (cf. Sharps & Martin, 2002, who focus on individual, not family-level, mindfulness/mindlessness). They may prefer to keep things that way. They may not be willing, or even able, to afford the costs (in terms of time, money, cognitive effort, inconvenience, confusion, or putting other things aside) of looking for and acquiring more comprehensive and more penetrating information (cf. Lawrence, 1999, pp. 28–32). But then they may not have a choice to know more than they know. The decision environment, the experts, the publicly available information, and the information provided by realtors and sellers, adoption workers, popular literature on parenting, retirement planning experts, and stockbrokers may provide only limited information. Thus, a family may be unaware of the existence of large and diverse amounts of relevant information, some of which could have a profound effect on their decision if they became aware of it.

People may also not know much that is relevant to decisions they must make that are of great importance because the available information is so overwhelming that they skim it, sample it, or ignore it. For example, if they Google "home buying" and have twelve million hits, they might decide not to use the information that Google offers. In fact, their obliviousness, in the sense of avoiding information overload and keeping information processing manageable, almost certainly facilitates their movement toward a decision in a relatively brief amount of time. An exhaustive examination of a substantial number of information sources and decision choices could leave a family with strong ambivalence about all options and with more to process than they possibly can, which could slow them down considerably or even prevent them from making a decision.

Perhaps the image of a couple or a family discussing or arguing about decision options is an image of each person advocating for her or his position. "Let's buy the house near the school." "No, let's buy the house that is more affordable." But from an information-processing and obliviousness perspective, it seems likely that, in some couples and families, decision discussions or arguments are often in part about limiting information. "Let's not consult with that expert." "Let's not look for anything more to read." "Let's decide by the end of the month on the basis of whatever we know then." "Let's stop researching and decide now."

In addition to moving toward a decision more easily when there is a limited amount of information, the qualities of the information people use can also block or facilitate forward progress. In particular, it may be easier to move forward if there is considerable obliviousness to undesirable aspects of what ends up being the family's choice and to desirable aspects of the choice alternatives that end up being rejected. In Festinger's theory of cognitive dissonance (1957), limiting information was seen as part of an individual's reaching comfort and commitment about a choice after it was made. But from the perspective of obliviousness in family decision making, a case can be made that obliviousness is integral to family decision making itself, that before a shared decision is made, family members are relatively oblivious to the undesirable aspects of the decision they are on their way to making and the desirable aspects of the options they are on their way to rejecting.

The argument made here about obliviousness and decisions is based on the idea that in many situations it seems better to make a decision than to be stuck for a long time while trying to resolve ambivalence, trying to find additional information that might help to inform or resolve a difficult decision, or trying to process more information than is possible to process. A family has to live somewhere; they cannot stall forever about picking a place to live. A couple who might or might not have a child has to decide while they are still young enough to have a child or adopt one. Starting a retirement savings plan when relatively young is preferable to reaching age sixty-five still looking for what would be clearly and unambiguously the right place to establish a retirement savings account.

If, in order to make a major decision, families must keep the decision process relatively simple, family decisions will have to be made with few explicit decision criteria in mind. To support and reinforce such a decision process, the larger systems will push families to focus on a small number of criteria. Presumably those criteria are not haphazardly arrived at but are important in ways that serve and make sense in larger systems. For example, if a family is going to have to decide whether to remove a relative from life supports, the medical and religious systems may push the family to use relatively few criteria. These institutions may focus family attention on those criteria that represent the moral, legal, and financial interests of the institutions. Let's say the criteria are "quality of life" and "what the person who is very ill would want." Then the larger systems will focus on developing advanced directives (so that a family can know what the person might want when terribly and incurably ill), on discussions with family members of what the very sick person would want, on what the person will be able to know and to do if continuing to live, and on cost in terms of physical pain, medical supports, and caregiving if the person were to continue to live. The focus does not then go to many other issues that could conceivably be relevant to the decision—for example, what precedent the decision will set in the family, the possibility that at a more elite medical institution the person's illness could be reversed, the possibility that there will be a great deal of horror and psychic pain (even if not communicated) for the ill person if removed from life supports, the financial interests of the hospital and health insurer in not continuing medical care, or the fact that there are ethical systems that would advocate never removing a person from life supports once the person is on them and unable to live without them.

Since the United States is dominated by corporations economically, politically, and in other ways, it would not be surprising that family decision-making obliviousness may often be (unintentionally and without awareness) in the service of corporate profits. If consumers are aware of the negatives about purchasing this car, that car, or any car, a piece of real estate, or anything else, and if they are aware of the negative sides of entering a particular career or entering an education program leading to that career, that may thwart corporate goals of selling products and filling jobs. Advertising does not emphasize the negatives about a product; companies do not emphasize the negatives about working for them, and lenders of financial aid for college students do not emphasize the negatives about educational paths leading to various occupations or about borrowing in order to finance education. The push is for what serves corporations. So obliviousness in families making decisions about a wide range of issues may be in important ways in service to and shaped by corporations. As part of that, the larger systems of corporations, corporate public relations, advertising, and the corporate media may generally work at keeping families oblivious to the negatives of whatever it is that families might do that serves corporations.

From another angle, family decision making is sometimes stymied by family members not compromising. If there seems to be no decision option that is acceptable to all the players, sometimes obliviousness works against the family members. In particular, when compromise seems impossible, there may be obliviousness to options that could make for a mutually satisfying decision. Even on a decision that seems impossible to compromise—for example, a couple for whom San Francisco is partner A's best job opportunity while New York is partner B's—there are options that make it possible for both partners to gain a great deal. Perhaps the key is not to be oblivious to alternative positions that could be satisfying to both partners (Fisher & Ury,

1981). For example, the partners could take turns, three years in partner A's best place, three years in partner B's. Or the two could try to negotiate high enough salaries so they can commute to see each other frequently, can support two places to live, and can afford the electronic arrangements so that they can interact visually on a daily basis while far apart.

Ethical Obliviousness

Perhaps related to issues of making decisions in a corporatist society, in making purchases many families are oblivious to the ethical implications of their product choices (Ehrich & Irwin, 2005). For example, they may not consider the record of a manufacturer with regard to the treatment of workers or polluting the environment. Although manufacturers and retailers may work at hiding such information, often information is available. But purchasers may not seek such information and may be reluctant to use it even if they discover it (Ehrich & Irwin, 2005). Using such information may complicate decision making, may block decision making, and may make people feel bad about a decision that might be the best choice on practical or aesthetic grounds.

Obliviousness is not infrequently considered a valid excuse in the United States for moral errors or lapses (cf. Dana, 2005). For example, Dana (2005) addressed ways in the United States that conflict of interest is considered by many to be more excusable and less likely to make very serious trouble if one argues that one was ignorant of the conflict of interest. Extending that to families, there is a sense that obliviousness protects a family, at least in some cultural contexts, from feeling that a decision was not moral and protects them from self-accusation and the accusation by others that they have done something immoral. "I didn't know that clothing made in Saipan was made in slave labor conditions." That excuse works much better, and is an honest claim, if one has worked at not knowing anything about the conditions under which the clothing family members purchased was made. So obliviousness is something of a defense against criticism of the morality of certain decisions.

To take another situation with moral elements, family members may prefer to remain ignorant about whether they are genetic carriers for certain incurable, late-onset diseases (Yaniv, Benador, & Sagi, 2004, whose research focuses on individuals). Information that a family member is genetically predisposed to such an illness could greatly complicate important family decisions, make the future more difficult to plan for, and complicate relationships in the family and with others. There are a number of ways to understand not wanting to know such genetic information, including that the reluctance to know is rooted in individual anxiety more than in family dynamics, but in a family where people share genes they may also share what Yaniv, Benador, and Sagi (2004) call "protective ignorance."

From another angle, a family's obliviousness is likely to reflect important values of the larger systems in which the family is embedded, and what the family does may reinforce those values. The larger system, it seems to me, values a degree of family obliviousness about the negative aspects of deciding to live the way the larger system values families living. So if corporations, the corporate media, and other entities with considerable power value a family's deciding to own and watch a television set or to have a telephone, it seems to me that the society values a degree of family obliviousness about the negative side of such ownership. Further, society deems it moral to go along with societal values and immoral to oppose them. So a family's obliviousness that is in accord with societal values can be understood as moral action. A family is a "good" family to the extent that it is oblivious in decision situations to lines of thought alternative to those laid out in the larger society—for example, oblivious to the negatives about deciding to own and use a television set or a telephone.

Family Obliviousness to Family Decision Process

In many families there may not be a great deal of awareness of how family decisions are made. Sometimes family members are not even aware that a decision is made until after the fact (Sillars & Kalbflesch, 1989). Or they may be unaware of whose influence and opinions count in what ways or whose views are ignored or discounted. They may be unaware of who has what kinds of information and who approaches the decision inquiry in what ways (cf. Hanks, 1993). They may be unaware of how much a decision is rational versus how much it is about this family member's desire to keep up with the neighbors or that family member's whim or anxiety. They may be unaware of how much the family decision process involves coalitions among family members.

Conceivably a family's awareness of its decision process would enable the family to make decisions with better information and better use of information and in a way that might be more fair to all family members. But just as shared family obliviousness about many attributes of choice alternatives may be useful in decision making, shared family obliviousness about the family's decision processes may also facilitate decision making. If family members become aware of their family decision processes, they may spend considerable time discussing the processes, and some family members might resist decisions because they do not accept some aspect of the process. Instead of decisions, a family could encounter long-term and perhaps irresolvable conflict about decision process.

Family Decisions that Maintain Shared Obliviousness

With obliviousness so important, it stands to reason that a family's decisions are not only about the decision issues but also about maintaining shared family obliviousness. To maintain shared obliviousness, they probably cannot say, "Let's remain clueless about this matter." Perhaps their decision process will be of a piece with their other processes for maintaining shared family obliviousness; that is, they will build on their very substantial resources of shared obliviousness. They will have the same blind spots, the same ways of discussing and not discussing in order to maintain obliviousness, the same limited ways of acquiring information they deem relevant to making a decision. In such situations, I assume a family's "decision" to remain oblivious may resemble something like a decision that certain kinds of information are all that is needed or that more information is not necessary or that one or a small number of criteria are all they need pay attention to (cf. Paolucci, Hall, & Axinn, 1977, pp. 116–17, citing Miller, 1965, p. 367). "Is the house we are looking at in the right location for us and affordable? Then let's buy it. Let's not investigate how good the kindergarten is at the local elementary school, who our child's playmates might be, or how comfortable this house is in other seasons of the year."

Families may also, in a sense, decide not to decide, by putting off decisions. They may put off decisions in order to avoid distress, guilt, conflict, painful consequences of choices, and other difficulties that could arise were they to make a decision about the issue in front of them. McCown, Johnson, and Carise (1993), for example, discussed a case where a family put off making the decision that the gangrenous leg of a family member had to be amputated. Alternatively, families may put off decisions because they are inclined to put off probing into an area of family obliviousness. For example, a family that has worked hard to be oblivious to the implications of their child's having some sort of attention deficit may avoid deciding to consult a physician about the problem, to read about it, or even to discuss whether what seems like a problem actually is a problem. And if they delay long enough for the child to be old enough to discuss these matters, the child may be drawn into the pattern of not talking about the attention deficit, remaining oblivious, and not deciding to consult a physician.

More broadly, a family's decision about, say, where to live, may be a decision to continue to be oblivious about many things that go on in society. For example, a white family's decision to buy a house in a well-off white suburb can include a decision to be oblivious to a great deal of racism in the United States and its effects on the families targeted by racism, oblivious to the ways that the well-off suburb has become well off through government processes that benefit white people disproportionately (Lipsitz, 2006, ch. 1), and how the family may be harmed by living in the well-off suburb. A family's housing decision can be in part a decision to surround themselves with neighbors who share the family's obliviousness and who will not threaten it, and an education decision can be a decision to send one's child to a school where family obliviousness will not be threatened. A decision to subscribe to a mainstream

daily newspaper and to make network news the focus of family news watching might be a decision to learn little about the profligacy of the very rich,
the power of lobbyists, and the centrality of calculations about increasing or
retaining political power in the making of a great deal of federal and state
legislation. A decision about how to vote may, similarly, be a decision that
works to maintain obliviousness by opting to elect or reelect officials who will
support the obliviousness of oneself and one's family.

Family Obliviousness to Decision Consequences

This chapter emphasizes that many decisions are made without much information or without much information of the sort that could lead to questioning
the decision that eventually is made. So as the decision is being made, a family
may be rather oblivious to the possible negative consequences of the decision.
A case can also be made that long after the decision has been made, when family members are in a position to perceive the consequences of their decisions,
they may still be oblivious. For example, a family may not pay attention to how
the choice of a house to buy led to their child's going to a terrible kindergarten.
They may have no difficulty recognizing that the kindergarten is terrible, but
their thoughts do not go to the housing decision. Their thoughts might focus
on the incompetence of a teacher, the lack of supervision by a principal, or
the failures of a school board to mandate effective teacher evaluation. So what
seems like awareness also is about obliviousness. And that obliviousness helps
the family to continue to be committed to their decisions and to their decision
processes. Too much awareness of decision failures or of unwelcome surprises
as a result of decision making could undermine commitments already made
and could raise questions that carry into the future about the family's decision processes. Too much awareness of decision failures and decision ignorance could raise uncomfortable questions about the information environment
provided by the information sources on which the family relies or the family's
information-seeking and information evaluation processes.

One way families can remain oblivious to decision consequences is to
continue to focus on the quality of the initial decision making. "If, at the time
we decided to institutionalize our profoundly retarded daughter, we consulted
the right expert and had good discussions, then we did the best we can do."
Continuing to feel that at the time the decision was made, "we did the best
we could," can keep a family out of continuous reevaluation of the decision
as they acquire new input, potentially every day, on whether the decision had
good consequences or not. And it can keep them from questioning their current and future decision processes.

From another angle, and adding to the discussion earlier in the chapter about the possible desirability of decision making being facilitated when

families have and only want a limited amount of information, perhaps all major decisions in life are made with substantial obliviousness to what the individuals and families involved are getting into. For example, the decision to have or adopt a baby, particularly a first child, is, I think, usually made by individuals and couples who lack full awareness of the demands and costs of parenting that child. My guess is that people do not have much awareness, when they decide to become parents, of how much it will cost them financially, how much sleep they will lose, how many times they will be intensely anxious concerning the child, what the risks are of serious medical problems for the child, and how much the couple's relationship will be changed by the child in ways that seem less than ideal. Having substantial areas of obliviousness in making major decisions may be crucial to people actually making those decisions. Perhaps far fewer people would marry, have children, move across country, take on a costly mortgage, and make other major life commitments if they were not rather oblivious to what they might be getting into. So arguably the continuation of the species and the functioning of the economy depend on a certain amount of obliviousness.

Obliviousness to decision consequences may also be about obliviousness to what goes on in the family. A family may maintain the system it has or move to a system that works better for some or all of its members as a result of making a specific decision, but the decision might have been very uncomfortable to make, or even impossible, if family members were not oblivious to those decision consequences. For example, family members might not so easily be able to buy a television set if they knew that having the television set would cut deeply into the amount of time they have to interact with each other, their sex life, their sleeping, or their relationships with kin and friends. Similarly, imagine a family that makes a commitment to belong to a religious congregation that takes up many hours of family time each week and places strong demands and limits on what the family can do. The family might not have been so willing to make the commitment if they had overcome their obliviousness to how much freedom they would give up in order to participate in the congregation. They might not so easily have made the decision if they could have known in advance that they would lose much of their freedom to allocate their time on certain days of the week, much of their freedom to engage in certain recreations, much of their freedom to have friendships outside of the congregation, and even much of their freedom to express certain feelings on certain matters. Perhaps in some ways the family wanted to escape from freedom (Fromm, 1941), but the decision to join the congregation might have been much more challenging if they had not been oblivious to the potential escape-from-freedom consequences of the decision.

6

Family System Responses to Threats to Obliviousness

A FAMILY MAY ORDINARILY BE rather effective at warding off threats to its shared obliviousness. Families typically have defenses well in place, priorities, distractions, closedness to hearing this or that, and so on. In response to a threat to shared obliviousness that could conceivably break through established lines of defense, a family may add elements that protect an area of obliviousness better, new barriers, circumventions, explanations for why things are the way they are, or aversions that make family members disinclined to open up an area of shared obliviousness. At times, an element of shared obliviousness may be replaced by awareness and knowledge, but a new wall may rise to keep the remaining shared obliviousness secure.

Some shared family obliviousness may be ended or reduced because the environment in which the family functions contains information that ends it. There are thousands of information sources that family members may brush up against each day. They turn on the radio or the television, pick up a newspaper, a book, or a magazine, overhear a conversation, or see certain sentences on a Web site, and perhaps then there is a challenge to shared family obliviousness. A family member who experiences problems because of family obliviousness to some matter may act to deal with the problems in ways that threaten family obliviousness. Threats to shared obliviousness also come from people who are not bound by the family's rule system—teachers, neighbors, physicians, members of the clergy, acquaintances, co-workers, friends, and so on. How is family obliviousness protected from all these potential threats?

Dealing with Family Members Who Might
Threaten Shared Obliviousness

Limiting New Experience

Perhaps the first line of defense in limiting threats to shared family obliviousness is that families have ways to limit the new experience of members.
For example, parents often have a role in choosing the preschools, elementary schools, high schools, and colleges children attend and perhaps even the
teachers they have within those schools. Some parents even choose to home
school, and that may be in part or entirely about limiting a child's exposure
to information.

Families may keep certain information sources out of the house and out
of the lives of family members. Usually when we think of family censorship
we think of families keeping young children from sexually explicit content
on television or in music lyrics, violence from these sources, or alternative
religious views. Presumably, the censorship is to protect the innocence of
children while they are in a vulnerable formative stage. But even in those
instances, and certainly with regard to more grown-up children and to adults,
there may be censorship that is in the service of broad areas of shared family
obliviousness. Adults want to protect their children from realities that are
inconsistent with family realities or from asking questions that would put
into the spotlight what the adults and children have so far ignored.

Often with shared obliviousness as an underlying force, family censorship may be passive, not active. If, for example, a family is oblivious to the
existence and experiences of people around the world who have been cap-
tured, imprisoned, tortured, and murdered by U.S. government intelligence
services, the family probably would not actively censor Web sites, news
magazines, and so on that report such information. They would probably
censor by never talking about the matter and having no interest in looking
for such material.

Cautiousness when using the Internet, when reading, or when in contact
with other information sources involves being selective about what one looks
for and the information one takes in. Cautiousness can be understood as a
defense against what we do not know is there that might impact obliviousness. It is not simply that in many families it is unthinkable to simultaneously
Google the words *CIA*, *prisoners*, and *torture*, but that family members may
be very cautious about Googling anything or opening any Web site or watching any television program that may have obliviousness-threatening information content.

Consider favorite television shows, favorite music, favorite musical art-
ists, favorite newspapers and magazines, and favorite Web sites. One way to

understand favorites is that they most entertain, please, and inform. Another way to understand favorites is that by focusing on favorites, family members guard against surprise information. They may have favorites in part because the content they offer is safely predictable. Favorites, in that sense, are not just pleasure giving and informing but are also protective of obliviousness.

What happens when family members accidentally encounter obliviousness-threatening information? There, for example, on the 10 p.m. local television news, at the place in the program that has usually been safely about a local fire, crime, or accident, is the beginning of a report about something obliviousness threatening—for example, about the large number of teenage prostitutes who come from middle-class families in the family's own neighborhood. What then? It would be consistent with arguments in this chapter for family members to do their best to minimize the threat. The channel is quickly changed, or the television set is muted or turned off, or someone among the family members watching the program starts speaking in a way that engages other family members in a distracting conversation. Perhaps members of some threatened families will telephone the television station to complain about the topic being covered. As for the family, there are probably ways to distance what was seen and heard. They do not watch and listen to most of the segment; they either do not talk about it or they find ways to discount it; and then subsequently they may well continue in their normal course of obliviousness to talk about the teen years as being about other matters than prostitution.

Obliviousness-maintaining censorship at the family level may also involve removing or never creating family records that could be threatening. Embarrassing or otherwise challenging photographs may be destroyed or never made, and old love letters from the wrong people may be destroyed. Only certain e-mail messages, school papers, medical reports, clippings, legal documents, receipts, and so on are kept. Perhaps this means that at one point somebody in the family meant to keep others oblivious to something, and may even have wanted to foster personal obliviousness. But then it is also possible that the censorship happens without awareness by the censor of the implications of the censorship for family awareness and obliviousness. The old love letters may be disposed of simply because the person disposing of them thinks they are no longer meaningful or wants to leave that chapter of her or his life behind.

Controls on Family Members with Little Power

Families have ways to silence (Zuk, 1965) or tune out anyone who has relatively little power in the family. There is a long history of women in a world of patriarchal families being silenced (e.g., DeFrancisco, 1991). In some families, a woman may not have the standing to speak about certain matters. If she speaks, her words may be ignored or scoffed at. In some families, women police themselves (e.g., Ussher, 2004), with their own feelings of guilt, self-blame, and shame that are silencing and make it easier for men to be oblivious to aspects of the experience of these women.

Then there is the silencing of children, the mentally ill, and those who are considered demented. Also, they may not be trusted to maintain family obliviousness. They may not be trusted to know the obliviousness rules or may be thought to lack the discipline to maintain those rules. Their observations, learning, questions, stories, and ignoring of injunctions not to talk about certain things can abruptly threaten to push the family system out of obliviousness about some matter and into uncomfortable or threatening awareness. Family systems may, however, be well defended against threats from children, the mentally ill, or the demented in part because other family members may segregate them, keeping them away from family discussions and conversations. Some children, mentally ill, and demented family members may be institutionalized at least part of the day, in schools, day care centers, day treatment programs, nursing homes, mental hospitals and the like. And they may be silenced in other ways. For example, some children are raised to be seen and not heard, or they are put in front of a television set or in a playground, ostensibly for their entertainment, but putting them there reduces their interaction with adults in the family and the possibility that they can threaten adult obliviousness. Children, the mentally ill, and the demented may be medicated in ways that silence them or keep them from saying some of what might threaten the obliviousness of others. And the mentally ill or the demented may, like children, be kept in front of television sets, which may or may not hold their attention but which might silence them.

There are in many cultures and families ways to discount what children say. Children are often blocked from knowing about certain matters (Goffman, 1959, pp. 212–13). Children are often ignored, or they are treated as cute, and their innocent questions that threaten obliviousness are laughed at rather than taken seriously. They are defined as immature and in need of education, and so their questions and comments that might threaten family obliviousness may be grounds for additional educational efforts to teach them the proper rules (privacy, secrets, rules of etiquette) for communicating in the family, including respecting obliviousness.

Redefining Information Deviants

People are sometimes defined as mentally ill on the basis of their deviance (Szasz, 1976), which includes their saying things that scandalize or threaten others. And they may scandalize or threaten others by saying things that

threaten family obliviousness or the family rules for maintaining obliviousness. Being defined as mentally ill can be a powerful and effective sanction.
Knowing that one would be sent to a psychiatrist, medicated, and less often
taken seriously because one kept asking about, say, who was hurt by how the
family acquired its wealth and property or why the family matriarch and
patriarch pretend to be loving when it is obvious that they hate each other,
would be a powerful force for choosing not to threaten family obliviousness.
Furthermore, in many cultures in the United States the mentally ill are often
dismissed as not making sense, not to be relied on, and perhaps even dangerous. That means that what someone who is defined as mentally ill says may be
dismissed as irrational, unconnected with reality, not to be trusted.

Similarly, with dementia, one way a person can come to be defined as
demented is to speak out in ways that threaten family obliviousness. When
Grandma starts speaking about, say, how Grandpa used to rape her, how
ugly a family member who everyone says is attractive is, or how empty family conversations are at family gatherings, that may be when family members
start to say to one another that she has become demented. That is when she
might be brought to a physician for a neurological examination and when
she might receive medications and be brought to classes that are supposed to
arrest the development of dementia. The label of "demented" discounts anything Grandma says. Then her threats to family obliviousness are defined as
sick, rather than veridical or insightful. All this may silence her. It certainly
becomes grounds for family members to treat what she says as nonsense.

So family members who might be most inclined to threaten shared family obliviousness may be marginalized, silenced, removed from contact with
the family, ignored, labeled as needing additional education, or otherwise
blocked or deterred from threatening family obliviousness. Despite this, the
threat to shared family obliviousness remains great in part because most people in most families function in more systems than that of a single specific
family. They may belong to more than one family. They may function in work
systems, organization systems, education systems, religious systems, friendship systems, and so on. Conceivably, each of these systems may bring things
to awareness that are in areas of shared obliviousness in a specific family. For
example, a minister may deliver a sermon about matters to which a family
has been oblivious. But perhaps people find ways to minimize intersystemic
challenges to shared obliviousness by choosing systems and specific locations
in systems that minimize threats. For example, if a white family has been
working at being oblivious to the racial system in the United States, they are
unlikely to join a black church.

Perhaps the threat of becoming marginalized, silenced, removed from
contact, ignored, labeled as needing additional education, or otherwise dealt
with in harsh ways will police most family members. They will not seek

out novel perspectives and information. And if they do by chance acquire some awareness that could conceivably threaten shared family obliviousness, the threat of what might happen to them if they speak up could keep them from acting to threaten the shared family obliviousness. The student who has learned in high school or college about feminist thought is not inclined to speak up about gender inequity in her or his family if gender inequity is an area of shared family obliviousness. Still, the potential that most family members may have to threaten some areas of family obliviousness probably keeps families vigilant to threats to obliviousness from inside the family.

Dissenters and Whistle-Blowers

The threat to shared family obliviousness from inside the family may also be framed as the challenge of dealing with dissenters and whistle-blowers. The treatment of dissenters and whistle-blowers in the larger society was discussed in chapter 3. Family systems may also have dissenters and whistle-blowers with regard to specific areas of shared family obliviousness, those who do not share family obliviousness to some matter and who are inclined to speak up about what they know. Families may also have those who speak out against certain ways in which shared obliviousness has been established or is maintained. Family dissenters or whistle-blowers may resent certain obliviousness rules, seeing them as inconsistent with other values of the family, as personally hurtful, or as hurtful to other family members. Some may find it trying to remain oblivious because they are curious about matters in a zone of obliviousness or they are connected through friendship, romance, marriage, or occupation to one or more persons whose experiences and knowledge lead them to speak about matters that are in the family's zone of obliviousness. Consider, for example, a white person from a white family whose closeness to a black person undermines the obliviousness to white privilege that she or he had been sharing with other family members. She or he may become a dissenter or whistle-blower in the family.

Family dissent and whistle-blowing are not easy. Dissenters and whistle-blowers may be marginalized, ignored, not taken seriously, and kept out of the center of the family. A dissenter or whistle-blower may be considered by family members not to be a principled person or one whose reality counts, but as someone who is rude and disrespectful or not to be trusted or a traitor. A dissenter or whistle-blower may be labeled as disloyal, acting out, adolescent, ill-mannered, crazy, or difficult, and that labeling marginalizes them and diminishes the power and influence of what they say. It also may discourage them from continuing on with their dissent or whistle-blowing, and so pejorative labeling is a family system defense of obliviousness (cf. Hopper, 1996). In a family where the stakes associated with some areas of shared obliviousness are high, a member who might threaten that obliviousness may not be allowed at the table for decision making, planning, or information exchanges.

In corporations and government, and quite possibly in families, whistle-blowing is often an act of a person who will not long be in the system. As is discussed in chapter 3, despite whistle-blowing protection laws, corporations and government agencies often find ways to eject a whistle-blower or to make things so difficult that the whistle-blower leaves. Families do not have whistle-blower protection laws, so a whistle-blower (for example, the person who says that Grandpa had sexually abused several children in the family) may not be listened to, may be treated as a liar, may no longer be invited to family gatherings, and perhaps may no longer be counted as a family member.

From another angle, corporations and government agencies often work hard to transform or limit the impact of what a dissenter or whistle-blower has to say. One way to do that is to blame the dissenter or whistle-blower for the problem the person is pointing out. Another is to blame specific insiders or outsiders who are convenient to scapegoat, but not the system, for the bad things that have happened. Extended to families, if, for example, a family whistle-blower or dissenter points out that the family has benefited for generations from a white privilege system that enriched it by oppressing blacks, Latinos, American Indians, immigrants from Asia, and so on, the family may defend itself the way corporations or government agencies often do. It may blame the dissenter or whistle-blower—for example, for distorting history. It may blame specific individuals (great-grandpa), outsiders (the railroad barons who created the land allocation system in the part of the United States where the family lives), or some other people (liberals who are giving a bad name to the hard-earned success of good people). In this example, the goal in defending the system may not only be to avoid feelings of guilt or feelings that what the family has is undeserved but also to defend what remains of the family obliviousness to its skin privilege.

Reactions to Those in the Family Who Bear Witness to What Others Ignore

There are those who bear witness to injustices, torture, murder, sexual abuse, and other dreadful matters. That is, they communicate to others their awareness of these horrible occurrences and their disapproval. For many who bear witness, it is a moral act, an act of speaking out against injustice and quite possibly a way to try to prevent future injustice.

Governments may try to disperse bearers of witness who gather in public places (Westerman, 1994)—for example, tear gassing a gathering of those who have come together in a city square to deliver testimony or to hear others do so. Because of government actions, those who would promote testimony about victims of, say, unjust government action may become new or renewed victims, but that may not stop them. Liberation theology speaks of "conscientization," the process of gaining consciousness of unjust political and social systems (Westerman, 1994). For some there are religious conscientization foundations for bearing witness and communicating that to others. Their testimony may be delivered to those who have engaged in and promoted the injustice, but often their testimony is intended for those who seem to have been oblivious (Westerman, 1994), for example, for U.S. citizens who have been oblivious to their government's involvement in training and supporting torturers and assassination squads in Latin America.

In families, the bearer of witness has no town square in which to speak and possibly no allies, but still there may be similarities with what Westerman wrote about. Other family members may try to stop the bearing of witness and may threaten the person with ostracism. But the bearer of witness may hope to conscientize others in the family to the unjust acts that have occurred and may have religious reasons for doing so. And the target for conscientization may be those who seem most oblivious and yet who have directly or indirectly been involved in making horrible things happen.

Realities That Cannot Be Ignored

Sometimes obliviousness cannot be maintained because a family is faced with realities too compelling to ignore. A family cannot continue to be oblivious to Grandmother's increasing dementia when she cannot find her way out of a bathroom stall or across the street to a neighbor's house. A family cannot ignore aging when someone becomes too feeble to get out of bed by herself. A family may not be able to ignore physical abuse when it occurs at a family gathering. This does not mean shared family obliviousness has failed and has ended. Shared family obliviousness may have held off awareness and realization for a long time. For years shared obliviousness may have kept family members comfortable. For years it may have saved family members from having to make difficult decisions or to change their thinking and behavior. And now, even in the face of realities too powerful to ignore, there may still be considerable shared family obliviousness. Yes, they can no longer be obliviousness to Grandmother's dementia, but they may be oblivious to how severe it is, how much it is incurable, how much it is permanent, how much worse it can get, how many people with dementia that severe have been institutionalized by family members, and the heritability of dementia. So even if realities can no longer be ignored, the months or years they were ignored still accomplished something in and for the system. And ending shared obliviousness about some matters still can leave shared obliviousness in place about many other matters.

People from Outside the Family System

Family systems theories address the boundedness of a family from outsiders (Rosenblatt, 1994). Families are rarely completely closed to the outside, but they typically have various kinds of barriers and limits to keep outsiders and outside information from altering the family system in undesirable ways. People entering a family system from other families and cultures may not know what the family system rules are about obliviousness, so they may unintentionally say or ask something that threatens to end an important obliviousness. For example, a visitor from an exotic culture, a new daughter-in-law, or a visiting social scientist might ask the members of a family why they watch so much television or why they claim to be ordinary people though they live in a house bigger than 99.99 percent of the houses that families around the world live in.

Xenophobia, ethnocentrism, homogamy, and other ways of being closed to outsiders can be understood in many ways. They could be about the discomfort of cross-cultural communication, a rejection of customs and beliefs different from one's own, a fear of one's children being corrupted, a lack of interest in people different from one, or a sense that the only people worth paying attention to are people who are more or less like oneself. Embedded in and hidden under all this could be a sense that it might be threatening to relate to people who cannot be counted on to respect shared family obliviousness and familiar obliviousness standards and rules. For example, part of the discomfort of cross-cultural communication could be that one has to be so much on alert to fend off questions and observations that threaten obliviousness, and once those things are said they might be hard to ignore. Part of the rejection of the customs and beliefs of other cultures could be that they could threaten obliviousness by making alternative realities seem sensible, by problematizing the realities one's family has been taking for granted, or by opening one to new paths for learning about the world.

Adults may fear that their children will be corrupted by outsiders because they see their children as ill-defended against "bad" influences. Included in such influences may be those that would make the children aware of something to which adults in the family are oblivious. The adults may not know what they are oblivious to, but they still may sense that they are. Furthermore, they can fear that a child does not have the obliviousness defenses that adults in the family have. And then once the curtain of obliviousness is lifted for a child, something could change in the child and in the family that could never be returned to the way it was.

More generally, unwillingness to interact with people from other cultures may be a recognition that there are fragilities in one's own family system involving shared family obliviousness, even if one does not know what the family is oblivious to or why the family defenses of that obliviousness might be fragile. If visitors from outside the family must be dealt with, they may be dealt with in ways that trivialize or minimize their threats to shared family obliviousness. For example, a visitor from another country may be invited to visit a family in the United States who will focus the conversation on food differences, clothing differences, or other areas that may seem safely distant from the most important areas of family obliviousness. More generally, there often seems to be a striking simplification of the culture of people who are welcomed for their cultural difference, so they are acknowledged as different but then silenced from speaking to most of the differences that are obvious to them (cf. Ortega, 2006).

Being a visitor from another culture also risks threats to obliviousness. For example, a family on vacation in the United States from an exotic culture risks contact with people in the United States who may raise questions or issues that threaten cherished areas of family obliviousness. So the visiting outsiders have incentives to simplify and trivialize the differences between cultures and perhaps to keep the contact with the U.S. situation very superficial (for example, sightseeing and visiting Disneyland). Sometimes all parties in intercultural contact may want to minimize the depth of the contact. For example, international festivals are often striking in how much participants from all cultures involved focus on dance, clothing, and food, and how much participants avoid talking about what might be deeper and more threatening issues.

In the United States, personal definitions of who is a family member often exclude in-laws and even one's spouse (Schneider, 1980). They are understood to be outsiders. By making them outsiders, people may increase the protection of their family-of-origin realties, rules, and obliviousness from challenge. So, in a sense, the in-laws and even spouses are treated like visitors from an exotic culture when it comes to certain areas of obliviousness. One does not bring up certain issues with them and keeps them away from certain topics. Any time the discussion comes close to an area of obliviousness, they may be distracted, derailed, or ignored. For example, if in one's family of origin there was considerable denial of death and obliviousness about an array of issues related to aging, ill health, and death, one may stay away from those issues in conversations with in-laws and even in conversations with one's spouse. In such instances one can say that it is personal obliviousness that is being protected, but the obliviousness of the individual might well be linked to ongoing shared obliviousness throughout the person's family of origin.

Obliviousness and Family Therapy

Fᴀᴍɪʟʏ ᴛʜᴇʀᴀᴘʏ ᴛʏᴘɪᴄᴀʟʟʏ ꜱᴛᴀʀᴛꜱ with a presenting problem, the issue that family members say concerns them and that they want to have help with. Shared family obliviousness is, probably, almost never the presenting problem. That is, families rarely come to a therapist saying that they collectively are oblivious to something and they want help overcoming that obliviousness. Good family therapy may successfully focus only on the presenting problem and be of great help to a family. But in some schools of family therapy, matters about which family members seem unconscious or unaware are often a central focus. And as therapy unfolds, a therapist who works from a framework that makes the unconscious, repression, or denial a matter for therapeutic attention may sense that there is some important area of obliviousness behind the presenting problem that could profitably be dealt with. Furthermore, as therapy goes on, a family is likely to bring up matters other than the presenting problem, which expands the range of possible areas of shared obliviousness that might, for therapists who work from frameworks that attend to what is below the surface, be dealt with therapeutically.

A therapist who chooses to explore an area of shared family obliviousness may have to do detective work, but often there are clues. There may be evidence of obliviousness rules in action—for example, avoiding certain topics, changing the subject, family members interrupting other family members in a way that seems to silence the interrupted family member. And then hidden underneath evidence that some family members do not understand aspects of the family past may be a pattern of shared obliviousness. For example, in families in which older family members were Holocaust survivors, younger family members may be committed to a heritage and a set of values that they do not understand, because the Holocaust roots of the heritage and values are shrouded by parental or grandparental secrecy and silence (Mor, 1990). Perhaps this is a different kind of shared obliviousness than shared obliviousness without anyone being aware, because the young people know and can say that they are oblivious to quite a lot; they just do not know what it is.

Sometimes there are signs that an area of potential obliviousness is important even if the family denies it or seems clueless about it. For example, the family's presenting problem might be that they do not spend enough quality time together. But underlying that, and not articulated by anyone in the family, might be a pattern of disrespectful communication, a system of unwillingness to compromise, or a family member's overuse of alcohol. Any of these matters may become evident to a therapist as the therapy progresses, and the therapist may have good reason to think that any of these matters has dynamic links to the presenting problem even if the family is oblivious to the matters or the links.

Family Obliviousness to Interpretations

Although there are contemporary models of couple or family therapy that have clear roots in psychoanalysis (e.g., Bergantino, 1997; Scharf & Scharf, 2005), much of couple and family therapy has left behind or even shunned ideas of psychodynamics and the value of analytic interpretation (Flaskas, 2003; Scharf & Scharf, 2004). Still, in major models of family therapy such as cognitive-behavioral, systemic, and structural there can be a sense that a goal of the therapy is not infrequently (and often not explicitly) to help family members deal with what they seem to be oblivious to or unconscious to (Crago, 1998). The word *oblivious* is perhaps never used in writings in these family therapy traditions, and therapists working with these family therapy models may rarely offer explicit interpretations. But the therapeutic questions, conversations, explorations of narratives, exercises, suggestions, comments, and homework may not infrequently be aimed at matters to which a family may seem to be oblivious.

When family members seem oblivious to something that might be usefully addressed by family therapists working with analytically derived models in which the unconscious, repression, and other aspects of psychodynamics are important or by family therapists working with other, very different kinds of models, what are the matters to which the family may be oblivious? Often, I think, the issues of obliviousness that come up are narrow but important areas of family interpretation of events. That is, the family members are aware of a great deal about something that is at the focus of the therapy, but their interpretations miss something that could be vitally important. For example, the therapy may focus on how much the teenage son of a divorced but still co-parenting couple is in trouble at school, at home, and in the community, and that might be where the therapy should focus. But the parents and the son may share obliviousness to the ways that the son's trouble holds the couple together or the extent to which, by the rules of his family, he has no other way to communicate about how sad and hopeless he feels because his parents could not get along well enough to stay together.

Intergenerational Issues

For some schools of family therapy, one extremely important area of shared family obliviousness is intergenerational relationships. These schools of family therapy, influenced by psychoanalytic notions of the unconscious, deal in part with what could be said to be family obliviousness to multigenerational patterns and relationships. From a multigenerational approach such as Bowen's (1978), Boszormenyi-Nagy's (Boszormenyi-Nagy & Spark, 1973), or Framo's (1981), there may be obliviousness about unfinished business in the family of origin. And that obliviousness means that the family is also unaware of how the presenting difficulties of the family can be interpreted in terms of the unfinished business. Imagine a family that knows that there is considerable tension about Dad's relationship with his mother and that Dad's mother is a source of tension and difficulty for the whole family. But they may be oblivious to how much it is Dad and the entire nuclear family holding certain ideas about loyalty to parents that is the key to the problem.

There may also be obliviousness about the similarity of patterns across generations. For example, imagine that among the adults in a family there is a pattern that also existed in their parental and grandparental generations. For example, perhaps in all three generations men made the choices for women about what reading matter came into the house and what was played on radio and television. Imagine, too, that a member of the youngest generation of adults and the person's partner are seeing a multigenerational therapist and the therapist leads the couple to become aware of the multigenerational pattern and of their previous obliviousness to that pattern. Perhaps simply uncovering the pattern and the obliviousness will have the effect of leading the couple to discuss the pattern for the first time, including the ways the pattern works for them or makes trouble for them. But perhaps instead the forces that made the couple oblivious in the first place to the pattern will continue to operate and the pattern will slide back into oblivion, with the couple, and quite possibly the therapy, focusing on other matters.

Loss Issues

Issues of obliviousness to interpretations also may show up when a family has trouble dealing with loss, or a particular loss. For example, consider the

Paul and Grosser (1991) analysis, cited in chapter 4, of family mourning. The Paul and Grosser approach to helping families who deny loss hinges, in a sense, on family members being oblivious to two interpretations of their not grieving a loss: (1) Not grieving means that they are denying their feelings; and (2) not grieving means that they are denying the significance of the loss. Similarly, Berkowitz (1977) wrote about helping families in which there was grief, depression, or other affects which the family had collusively denied, minimized, or disavowed. As Berkowitz wrote about these matters, families are aware of their overtly displayed emotionality, but they are oblivious (my word) to the interpretation that their emotionality on the surface masks deeper emotionality that is different from what is on the surface, and that underlying emotionality is disclaimed. Or, to take another kind of example, families in which sexual abuse is occurring may be aware of the sexual abuse but may deny the impact of what is going on or the duty of all family members to behave responsibly with regard to sexual abuse (Hoke, Sykes, & Winn, 1989).

Obliviousness to What Is Positive

The examples of family obliviousness to interpretations offered so far in this chapter have focused on family obliviousness to what could be understood as negative interpretations of what underlies their problems. The interpretations are "negative" in the sense of identifying something that family members might think would make them look like a less good family in the eyes of many people. Examples of that kind of obliviousness include obliviousness to conflicts or to blocks to emotional process. But families may also have their reasons to be oblivious to positive interpretations of what is going on. They may, for example, be resistant to knowing that their everyday interactions, which they take for granted, involve what therapists would consider strengths. Parents who are concerned about their children may be resistant to knowing that their therapist interprets some of what their children do as mature, self-assured, helpful to others, respectful, and competent. A couple who is thinking of divorcing may be resistant to knowing that there are ways to understand their relationship as good, healthy, decent, wholesome, caring, and supportive. These more "positive" interpretations may be resisted because they threaten areas of obliviousness that have become central to couple and family realities. For example, knowing that a relationship or a child has considerable strengths means that a couple or family has to abandon the simple idea that the relationship or the child is all bad. And that may mean there is much more for them to know and understand and that they will not be able to make simple negative statements. It will also mean that they do not have a clear-cut, simple, negative set of evidence on which to base decisions about what to do.

Family Obliviousness and Resistance to Family Therapy

Some families resist coming to family therapy. They do not come to family therapy because they deny that they have problems, that their problems are as serious as they are, or that they can receive help with their problems (Overton, 1994). One can take any of those foundations for resistance to going to therapy as being about shared family obliviousness—for example, shared obliviousness about the existence of their shared family problems, the seriousness of those problems, the potential value of therapeutic help, or the possible limitations of their own efforts to solve their problems.

Family therapists are at times confronted with what some call "treatment resistant" families (e.g., Anderson & Stewart, 1983; McCown & Johnson, 1993). Family members may come to therapy, but not all of them want to be there. If the therapist assigns homework, some or all family members do not do it. Family members say they want to change, but then they seem to fight change, or individual family members want other members of the family to change, rather than themselves. Family members say they want the therapist's help, but then they act as though they do not want to allow the therapist to influence them. That is, they may be unwilling to go along with what the therapist says the family might try to do in order to deal with their problems, or they may be in some other way unwilling to change. And they may drop out of therapy if it moves closer to actually helping the family members to address their shared issues or if the therapy becomes threatening to cherished aspects of the status quo. Contemporary views of treatment resistance often define what seems to be client resistance as a sign that the therapist has failed to be wise, flexible, and competent, or they locate the resistance in the therapist-client system. In either case, these contemporary views can be said to empower the therapist and head off the therapist's blaming the clients for problems in the therapy. However, it may be useful to explore the ways that a client family, no matter what the therapist does or who the therapist is, may resist family therapy. Underlying any expression of treatment resistance may be shared family obliviousness, family members not being aware of what is going on in the family or what in the family makes them react as they do to the family therapy.

One can take the therapeutic efforts in many schools of family therapy as in part implicitly addressing areas of family obliviousness that can underlie family resistance to change. For example, Anderson and Stewart (1983, p. 18) wrote about structural family therapy (e.g., Minuchin & Fishman, 1981) as employing strategies to alter family rules that make trouble, and those rules can be seen as outside of family awareness, that is, matters of shared family obliviousness. Or, to take another example, according to Lantz (1992), in family logotherapy families that advance toward an awareness of meaning and purpose in family life will ordinarily resist the vulnerability and responsibility that comes with experiencing love and meaning in family relationships. The resistance, as Lantz saw it, involved blocks to actually knowing the others in the family and to taking responsibility in relationship to them. One could make a case for seeing those blocks as a form of striving to maintain individual obliviousness about others in the family and about what would be involved in a genuinely loving relationship with them. And with a family whose members collectively are blocked in that way, it might be argued that there could be shared family obliviousness to resistance in the family to the vulnerability and responsibility of greater intimacy.

Therapy Dealing with Processes of Obliviousness Maintenance

Reflecting discussions throughout this book, shared family obliviousness could be an issue in family therapy not because of the content of the obliviousness but because the ways that obliviousness is policed and maintained may create family difficulties. For example, family trouble may come from the family rules, the family sanctions for rule violations, the family closedness to certain kinds of information, the ways that some family members may not answer the questions of others or may change the subject, family secrets and myths, or how family members with power maintain that power. In these cases, the challenge in therapy is quite possibly to address how the family deals with curiosity, questioning, difference, disagreement, and knowledge.

From another angle on obliviousness maintenance, sometimes to understand what has brought a family to therapy a therapist may find it useful to understand what is not being attended to by the family that might be a crucial underlay for why the family has its problems. The therapist might explore with the family, its priority setting, its socialization of children, the extent family members keep very busy, family avoidance of contact with certain others, the possibility of specific traumas in the family's past, and other matters that might be linked to shared family obliviousness. Searching for what the family is apparently oblivious to may reveal to the therapist and perhaps the family that what the family does is undergirded and perhaps driven by efforts to stay away from things that the family is working at being oblivious to. For example, keeping busy may be a powerful way for the family to avoid dealing with differences or with great emotional pain.

For some families and some areas of shared family obliviousness, the family may be able to carry out its own exploration of shared family obliviousness. For example, in therapeutic work with parents of infants where there is what might be called an infant mental health issue, it can be helpful to engage the parents in self-observation about how they interact with the infant and with each other in relationship to the infant (McDonough, 2000).

Therapy Dealing with the Impact of Context on the Family

Family obliviousness is to some extent a defense against potential threats from the world around the family. Shared obliviousness may enable a family to function well enough by certain standards in the face of threats from global warming, globalization of the economy, economic changes that threaten the family's middle-class way of life or well-being, and other matters that would alarm them, make them anxious, or otherwise create difficulty for them. But if family members do not effectively tune out something that is a source of such great anxiety that it makes great difficulty for them, it may be appropriate to address the issue in family therapy. (See, for example, Simon, 1990, who focused on how anxiety over the threat of nuclear war could be an issue for families.) A first step in deciding whether a potentially problematic issue that a family is not talking about should become an issue in therapy sessions might be for the therapist to bring it up. If, as is common in therapy, therapists do not bring up possibly threatening issues in the larger context, families are unlikely to find help dealing with such issues in the therapy (DeMuth, 1994; Simon, 1990). A question about fear of global warming, nuclear war, or the changing economy might elicit an answer from a family with the presenting problem of parent-teenager difficulties that, yes, the teenager sometime says something such as, "What's the point of doing homework or planning for the future when the planet will be uninhabitable in the future?" (Reusser & Murphy, 1990). Then there is something to explore with the family.

One question family members might raise if a threat in the larger society is real for them might be what the family can do in the face of events controlled by powerful and distant others (Simon, 1990). And then, as with other kinds of issues where families feel helpless to deal with what is very difficult for them, a goal of the therapy might be empowerment. Defining the issue as beyond the family to address is not nearly as helpful as finding ways for the family to come to a sense that they can do something to contend against the problem (DeMuth, 1994). For example, the family might be helped to define political activism, public witnessing, or letter-writing campaigns as constructive ways to address certain contextual problems (DeMuth, 1994; Simon, 1990).

From another angle, shared obliviousness is costly to a family if it blocks the family from possibly successful efforts to make needed and desirable changes in systems with which the family is in contact. For example, if a family is oblivious to how they could act to make the school their children attend better, how they could act to make a road they travel safer, or how they could act to help bring the community drinking water purification system up to more desirable standards, the family's obliviousness hurts it.

Obliviousness in the Social Systems of Family Therapy

Therapy clinics, professional organizations of therapists, therapist education programs, state boards of therapy, and therapist consultation groups, like all systems, have their areas of shared obliviousness (Cooklin & Gorell Barnes, 1993). For example, there have been times and places where therapists seemed oblivious to family incest and to issues concerning race and ethnicity (Cooklin & Gorell Barnes, 1993). The frequent focus in some approaches to family therapy on interaction patterns and family structure may represent a culturally based discomfort with internal or intrapsychic events by members of the family being seen and by the therapist (Stein, 1985). A family therapy focus on behavior and externalities may be seen as part of the problem for a family, not part of the solution (Stein, 1985), because it is based on obliviousness to inner meanings, inner feelings, and quite possibly what the family and their therapist would rather not know. Therapists who rely on office visits by client families may typically be oblivious to important aspects of the home life of clients. Education in family therapy theory focuses on what it focuses on—for example, coalitions, triangles, communication problems—and may be oblivious to much that is of great importance to working with families that family therapy theories do not address—for example, couple and family sleeping patterns, home cleanliness, how a family uses the spaces in which it lives, family record keeping, the objects and clothing a family has stored, or what the neighbors think about a family. Therapy systems may also have a degree of obliviousness to the inevitability of problems and conflict in individual and family life (Gerhart & McCollum, 2007). According to Gerhart and McCollum (2007), mental health fields may perpetuate a myth of a problem-free life with concepts such as "mental health" and the implication that every problem has a solution. Whatever the areas of obliviousness of therapists in general or of specific therapists, it is possible to reduce obliviousness in areas that are important in therapy with some families. Some therapists, for example, tell of having increased their awareness about gender issues in ways that have made them more effective with certain client families (e.g., Brooks, 2001). And the history of family therapy is to some extent a history of learning to see and know aspects of family life that previously were matters of considerable, if not total, obliviousness.

Therapy may also model obliviousness. The therapist's disinterest in hearing the details about a couple's latest squabble or the never-forgotten insult of eighteen years ago, the therapist's focus on one or a small number of specific issues, the therapist's lack of curiosity about most of what members of a client family think or do, all model for a client family a way of defining and solving problems that involves a great deal of obliviousness. This may in itself be therapeutic, teaching a family that a system with a great deal of obliviousness may function very well and may meet the highest standards of a respected expert. For example, for the family in which people are miserable as they haggle about a million different apparently minor matters, the incisiveness of therapist focus is a powerful message to move to a pattern where family members do not pay attention to most of the details that have been the focus of their squabbles. Instead, they had best figure out what is most important and focus on that.

From another angle, just as families may be oblivious to much in the larger society that affects them, so may therapy. Perhaps that is often no problem in therapy, but sometimes therapy may be oblivious to matters of great importance. Imagine living in a society in which violence is endemic. There are political assassinations. Street people are routinely murdered by the police. Drug gangs carry out bloody warfare. The police and various private armies routinely torture and "disappear" enemies of those with power in the society. Investigative reporters and newspaper editors may be murdered or must flee the country to stay alive. Labor organizers are routinely murdered. In such a society, family therapy that does not explore the family's awareness of the violence and their possible vulnerability to or involvement in it and that does not probe for possible family obliviousness to the violence would be doing the family a disservice. The fear, the losses, and perhaps violence itself almost certainly come home to that family in various ways, both at the individual level (in terms of generalized anxiety, if nothing else) and at the family level (perhaps in terms of tense and difficult interactions, with the threat of violence and actual violence part of the family's home life). Astute therapists may be aware of the violence or other contextual factors and bring them up in therapy with families. Therapy that is oblivious to contextual factors may be much less helpful than it could be.

What are the larger system characteristics that are widespread in the United States that family therapy might best not be oblivious to? It might be the high level of competition and the sense that there are some winners but many losers. It might be the consumerism and a focus on material goods. It might be the emphasis on youth. It might be the lack of a long-term view of the future. It might be dishonest and/or misleading communication in government, business, advertising, politics, and elsewhere. A therapist's decision that family therapy should deal only with a family's interior and not with their context is a political act and also, quite possibly, an act of neglect (Imber-Black, 1990). It means that the therapy may stay away from issues that have enormous impact on the family. Examples include the ways that social institutions push a model on the family of me-first competition with arrogantly proud winners and shamed losers, or the ways that many social institutions dealing with parents foster the notion that single parenting is not as good as dual parenting. Or, to take another example, a therapy system that works with black families that was oblivious to racial history and to ongoing racism and white privilege would probably be relatively unhelpful (Pinder-hughes, 1990). A therapy system that is oblivious to broader societal issues is in a sense allying itself with the forces in the larger society that make life difficult for families and that foster family obliviousness to the power and malevolence of these forces (cf. Imber-Black, 1990).

Shared Obliviousness in the Therapist-Client System

In therapy, the therapist and clients together create a system. Like any system, it will have its areas of shared obliviousness, and must have them. Where does the obliviousness come from? Presumably both sides import some obliviousness into the shared system. A therapist works as a professional at remaining oblivi-ous to whatever seems irrelevant to effective help with client family issues—fo-cusing instead on what is of interest from the viewpoints of the therapy models the therapist works with. Plus, the therapist may bring obliviousness from her or his family of origin and current family to the therapy. If, for example, in the therapist's own family, people do not talk with others about their physical health history, possibly that will be one of the obliviousness rules the therapist will import into the therapy (while oblivious to the importing).

Clients will bring their family system obliviousness into the therapy. For therapists working within certain models of how to do therapy, that oblivi-ousness might be grist for the therapeutic mill. For other therapists, the tech-niques they use may make attention to most or even all of a family's areas of obliviousness irrelevant to therapy. But therapists have to join with a family in some regards in order to gain their trust and to help them stay with the therapy long enough for the therapy to make a difference. It can be a chal-lenge to a therapist not to join so well with a client family as to buy into the family's obliviousness about certain matters, and families may withdraw from therapy if a therapist threatens to breach a family wall of silence and oblivi-ousness (Bernstein et al., 1989). In fact, there is reason to believe that not infrequently therapists working with client families who have experienced trauma carry out the therapy in ways that replicate the client family's silence and secrecy about the trauma (Abrams, 1999; Bernstein et al., 1989). At an extreme, a therapist might not even ask about a major trauma in the experi-ence of a client family whose everyday life may be pervaded by the trauma and who have a system of silence and obliviousness about it (Lang, 1996). Of course, usually therapists working from many different therapy frameworks and personal backgrounds readily explore issues of trauma with client fami-lies, but sometimes that does not happen.

As the therapist and client family create their shared system, part of what goes on is the mutual (and typically covert) negotiation of what not to talk about or be aware of. Some of that will be given by the culture of the participants, and particularly if they share a culture it may be easy to maintain the culture's areas of obliviousness. For example, white clients with a white therapist may form a system that is oblivious to white racial privilege or to the effects of discrimination on nonwhite families. This suggests a potentially very important question: What areas of obliviousness in the culture of most therapists and their client families might defeat therapy with those client families, limit it, or make it very difficult? One possible place to look is in the literature on the cultures of most therapists—perhaps U.S. cultural studies or the anthropology of the United States. If, for example, the therapist and the client family are embedded in a culture that values progress and sees individual dispositions as crucial to progress, one can wonder whether obliviousness to values other than progress (acceptance, say, or understanding) and other than individual change (coordinated family change, say) may limit therapeutic possibilities. And those kinds of shared obliviousness could also limit the possibility of insight into the power of what goes on in the larger society to drive what goes on in the family.

Family Therapy that Reduces Family Obliviousness

Arguably most families who go to family therapy are oblivious to important matters linked to the problems that bring them to therapy. They might be oblivious to how their own patterns of interacting get them into trouble, or even to the patterns themselves. They might be oblivious to a problem habit. They might be oblivious to how they are blocked from dealing with whatever their presenting problem is. There is obliviousness of all sorts that comes with a family to family therapy. It is often an important part of the magic of family therapy when it works that it directly or indirectly helps a family to address some sort of shared family obliviousness.

Perhaps most often family therapy that reduces obliviousness does not affect most areas of shared family obliviousness and does not have much of an impact on the areas of shared obliviousness that it does address. But perhaps occasionally changing one small area of obliviousness might lead to the undermining of a much broader area of obliviousness. Then there is the potential for information overload. Families can only handle so much information. So now that the family is no longer oblivious to Mom's addiction, to matters of addiction in general, and to addiction in the family context, they tune in on and talk about addiction much of the time. They read about addiction. They make friends in the recovery community. They attend workshops about addiction. And at the same time, to avoid information overload, they draw away from other things that had been important in their life. They might, for example, pull back from their religious community and lose track of the lives of people in that community and of the events and the politics of that community.

However, it is not necessarily that simple. Just as individuals differ in the span of information they seem to be able to process, assimilate, and use, so families are likely to differ in how much information they can handle. Staying with the example of families on the path to reducing obliviousness to addiction, some families may be able to handle a great reduction in obliviousness in that area without needing to become more oblivious in other areas. They have the capacity to add complexity and can pay attention to a substantial range of matters at more or less the same time (relationships, conversation topics, concerns, and so on). Other families can only handle a reduction in obliviousness about addiction if they give up a great deal in other areas. With a family that will have to add new areas of obliviousness in order to deal with addiction, a therapist might conceivably have to work with the family, as their attention to matters of addiction increases, to help them consider what they might give up, what activities, interests, pastimes, and commitments. However, successful therapy with a family struggling with an addiction or other problem may free up additional space in their lives for information, interactions, learning, new commitments, and so on. If Mom is no longer going to the local bar, if she becomes more available to do a fair share of family tasks, if family members no longer spend so much time struggling with the problems her addiction has created, that may free them to pay attention to more than they could before.

Why Therapists Might Gain from Attending to Obliviousness

This chapter and the entire book suggest ways that a practitioner who wants to understand and help a family might find it useful to think in terms of family obliviousness. Like any conceptualization of families, society, and helping relationships thinking in terms of shared family obliviousness can help in dealing with some issues and not others and can help when therapeutic work with a particular family seems to reach a stuck point and other approaches seem not to work or be workable. One advantage of thinking in terms of shared family obliviousness is that it is always present; families are always oblivious to vast amounts of information about what is going on around them and in the family. Although, as this book points out, there are many ways that obliviousness may be resistant to change, may be well guarded, and may even be valuable, family members may delight in certain discoveries about their own obliviousness. They may find themselves interesting, and thus be pleased with new information about themselves. They may find that a new awareness opens new lines of discovery and moves them out of stuck places they have not liked. Some breaches of obliviousness can make great trouble for individual family members or for a family as a whole, so a therapist must be judicious in moving forward with any possible breach of obliviousness. But that moving forward may not do harm because obliviousness is often well defended, and it may produce new awareness, learning, discoveries, insights, and patterns of relating that might be quite beneficial to a family.

8

Researching Shared
Family Obliviousness

Rᴇsᴇᴀʀᴄʜ ᴀʙᴏᴜᴛ ᴘᴇᴏᴘʟᴇ's ᴀᴡᴀʀᴇɴᴇss and what they attend to and know typically focuses on the individual, not social systems. Even research exploring awareness, attention, and knowledge in the family context typically draws on data gathered from individuals, not families. So research on family obliviousness, which seems to me to require data from family members together, not individual family members, would be a break from standard research approaches.

The Challenge of Researching
What People Seem Not to Think About

Researching family obliviousness would be challenging. How can an observer recognize that a family is oblivious to something? Much social research relies on self-report from the people studied, but if they are oblivious to something they have nothing to report. How does one research what is in people's awareness or not if they do not say anything one way or another? How does one interpret with confidence what it means that the members of a family do not talk about a certain matter?

Ethical Problems

There are ethical problems in asking a family about something that they collectively have been oblivious to. If they are oblivious to something, that obliviousness may be at the core of their values or their family functioning.

Questioning that breaches their obliviousness can be seen as an ethical violation, pushing them to be aware of and know about what their values have kept them away from or what has enabled them to function as well as they have as a family. Questioning them can open them to very uncomfortable feelings of powerlessness (Hopper, 1996) and can undermine their sense of themselves as moral beings (Hopper, 1996).

What is a researcher to do who believes it is unethical to seek self-report from families about matters to which they might be oblivious? The researcher may try to infer family obliviousness from what is not said and from what can be observed in the family, but that is epistemologically risky. Family members may be quite aware of something but not give off information about the matter in what they say or do. They may even give off information that indicates they are aware, but it could be "coded" in a way that the researcher does not understand. For example, the family is aware of global warming and has even replaced a number of high wattage light bulbs in their home with lower wattage light bulbs, but if the researcher does not know of the bulb replacement, the researcher may have no basis for picking up on the family's awareness of global warming through what they have done with lighting.

Questioning family members or writing about them in ways that could disclose information about any family member to any other family member is ethically risky (Rosenblatt, 1999). What if one family member learns something about the obliviousness of another family member that embarrasses or shames the one in relationship to the other or that sets off family resentment or conflict? Family member A's learning that family member B has knowledge that A did not want B to have can create family difficulty. Even though we might have reason to think that family members share obliviousness about some matter, probing for that could well come up against the internal diversity that all families have, and inadvertently revealing that diversity to family members could create difficulties among family members.

What is a researcher to do in the face of the risks of family difficulty if what one family member knows is disclosed to another? One possibility is that the researcher may, for ethical reasons, choose not to ask about certain matters or not to write about them. But the researcher may be able to head off these difficulties by interviewing family members separately and by writing in ways that omit or disguise what might create problems in the families who are researched.

Family Obliviousness Inferred from Expert Opinion

One way to decide that a family shares obliviousness in some area is to compare the family's level of awareness and information in the area with the awareness and information of a knowledgeable source. Illustrating the gain from such comparisons, Pollner and McDonald-Wikler (1985) wrote about a family that claimed that a five-and-one-half-year–old-child who was not talk-ing was actually very bright. Was the family oblivious to the child's language disability? The child was evaluated through standard methods of assessing child ability and found to be severely retarded. On the face of it, the research-ers were able to show that the family was oblivious to the child's mental retar-dation. It is possible that the tests and testers were somehow oblivious and the child's family members were not, but Pollner and McDonald-Wikler give credence to the tests and to unanimous agreement among the evaluation staff. Probably Pollner and McDonald-Wikler would also have said, if asked, that the testing approaches used had been determined over a number of years and after considerable research and practice to be valid at assessing child ability. But that could conceivably mean that there has been long-term professional obliviousness to the limitations of the tests and the test validation process. Assessment instruments are generally validated by comparing the results they produce to information from other measurements that are considered valid. If there is agreement among the different sources of assessment, that is taken as validation. But it is always possible that the different approaches share oblivi-ousness and hence error. Related to this, Campbell and Fiske (1959) showed how different measures of supposedly the same trait or disposition may be contaminated by shared methods variance. So the agreement between the measures may be partly, primarily, or even only because they share methods variance (that is, they share error variance), not because they validly assess something. Still, there is a great deal to be said for using outside experts and methods of evaluation in deciding whether certain kinds of obliviousness are present in a family. I would not fault Pollner and McDonald-Wikler.

Sometimes expert observer realities are so overwhelmingly matters of agreement that there is almost nobody who would challenge those realities. For example, in the literature on *folie à deux*, *folie à trois*, and *folie à famille*, delusions shared by couples, trios, or families, there are reports of delusions that seem to be so disconnected with consensual reality that virtually every observer outside the couple or family would agree that delusion is present. Consider a family in which every member was sure that a woman in the family was shrinking (Ropohl, Elstner, Hensen, & Harsch, 2005). Despite the family's beliefs, the researchers made measurements and said that their measurements showed that she was not shrinking. It would seem to be the extreme of casuistry to question the medical authorities who published the paper. Surely they are right, that the woman was not shrinking. It seems such a simple matter to measure whether a person is shrinking or not. But there is always that small doubt about the possibilities of researcher obliviousness. For example, could the woman and her family have meant something dif-ferent than what the researchers thought they meant about her shrinking? If so, the researchers were measuring the wrong thing. I am impressed that

Ropohl and colleagues made measurements, but from the viewpoint of studying obliviousness, it would be remiss not to raise questions about conceivable, however unlikely, researcher obliviousness.

How Much Awareness Is Enough?

Another challenge in researching family obliviousness is the question of how much awareness and knowledge by the people in a family system is enough for us to say that they are not oblivious. Family members may share a degree of awareness about an issue, but have only partial knowledge or only a fraction of the knowledge they might need in order to deal effectively with the issue. For example, perhaps everyone in a family is aware that Mom drinks a great deal of alcohol and often is wobbly. But that is not the same as their realizing that she is drunk every evening, that she spends large amounts of money on alcohol, that she is addicted to alcohol, that her health has been seriously compromised by her drinking, and that her drinking has cut her off from friends and led her to lose her job. How does a researcher decide how much awareness or obliviousness is enough for a family to be counted as aware or oblivious about a matter? The researcher may well need to establish criteria of "good enough" family knowledge, in general or on a specific issue of interest. But there are ethical issues in making such judgments. These include the ethics of judging on the basis of imperfect knowing; the family may be much more aware than the researcher can know. The ethical issues also include the ethics of judging when the researcher is embedded in oblivious systems (see below in this chapter). The ethical issues also include the ethics of working from a researcher's perspective, as opposed to the family's perspective. The researcher could decide that the family is rather oblivious, but from the family's perspective the awareness they have may be perfect, given their values and understandings.

Then there is the related issue of family diversity. On many matters, family members will differ in degrees of knowing. How much awareness is enough when the members of a family range in awareness of some matter, from, say, partial or more or less full awareness to what seems to be complete obliviousness? The researcher will have to decide, presumably on the basis of the researcher's conception of shared obliviousness with regard to the issues being explored. Perhaps the most important thing is not to presume that any single family member knows what other family members know or do not know but to evaluate family obliviousness by gathering information from all family members.

Pseudo-Obliviousness

There is also the challenge of deciding whether what looks like obliviousness is that and not something else. People may choose to act as if they know nothing when they know quite a bit. The choice to act as if they know nothing may arise from fear or anxiety about the consequences of appearing to know something when they think others might punish them, feel unhappy with their knowing, or otherwise make difficulties for them because of their knowing. They may act as if they know nothing out of respect for others who might be embarrassed or shamed by what they know. They may act as if they know nothing because they do not want to deal with the challenges of doing something based on their knowledge. They may act as if they know nothing in order to continue to have access to information that would be closed off to them if it was revealed that they know something. And this is only a sampling of all the reasons people might have to hide, mask, or not disclose their awareness of some matter. Moreover, a family may be oblivious to something at the moment, because their attention is directed elsewhere. But as soon as their attention is directed at the matter to which they may seem to be oblivious, they may show that they are quite knowledgeable. For example, a family may seem to be oblivious to the fact that a substantial amount of the property tax they pay arises from their city government's using tax increment financing to subsidize a surfeit of businesses in their city. The family may not say anything about it even if one asks them what they think about living in their city. They may talk about the location, the parks, the schools, access to public transportation, and the quality of municipal services. So they might be taken to be oblivious to how tax increment financing is affecting their property taxes. But if one asks more directly about tax increment financing in their city, they may talk at length about the history and politics of that financing, and its impact on their taxes.

Knowledge Created as a Research Artifact

Questioning families about something they have not been aware of could create an awareness that they could then report on, so then one would be studying an artifact of one's own questioning. It could be a challenge in assessing people's knowledge about anything that the assessment can create knowledge. It could be particularly challenging in assessing family obliviousness to find ways to probe the non-knowing and nonawareness without creating some inkling of knowing and awareness.

To Perceive the System's Obliviousness, Someone Must Not Be Oblivious

Obliviousness is always about some people knowing and others not. If everyone were oblivious, there would be no way to know that there was obliviousness. Perhaps it is an obvious point and one that has been implied throughout this chapter, but it would be nearly impossible to research obliviousness shared throughout a family if there were not some way, independent of the obliviousness of the family studied, to know that there was information to which they were oblivious. So research about family obliviousness always is in part about observer awareness.

Among family members too, if someone is to perceive obliviousness in the family, that person must be aware. So in analysis of family systems in which there is both awareness and obliviousness, the systems analysis can be characterized in terms of where the obliviousness is located. Sometimes those who are oblivious are at the heart of a family system's problem, and those in the family who are not oblivious are struggling because of the obliviousness (imagine a family with an alcoholic father who is oblivious to his alcohol problem). Sometimes the people with power in a family are oblivious and those without power are aware (for example, farm or business families in which a scapegoated daughter-in-law understands the system of scapegoating, while those who scapegoat and who control the family enterprise and the family reward system do not). Sometimes the people with power in a family are aware and those without power are oblivious (for example, when family secrets are kept by powerful family members).

From the viewpoint of Maturana (1987), there is nothing remarkable or paradoxical about saying that someone must not be oblivious in order to recognize obliviousness. Maturana asserted that everything that is said is said by an observer. There cannot be assertions about what goes on anywhere unless the person making the assertions is an observer whose standpoint is located outside of what is being observed. A practical research implication of this viewpoint is that if a researcher is oblivious, unless there is someone in a family who can reveal what the family is oblivious to, the researcher is not going to learn about the family's obliviousness. If a researcher is oblivious to what a family is oblivious to (for example, if researcher and family members are oblivious to how the decline in the value of the dollar is affecting families), research focused on the experience of being in the family will turn up nothing about the matter (for example, the impact of the dollar's decline).

How can a researcher know that some issue exists about which a family is oblivious? Since oblivious families cannot report their obliviousness, a researcher has to have some kind of awareness and knowledge independent of that of the families studied in order to know that something exists about which the families seem oblivious. The researcher has to have some kind of independent learning. For example, if a researcher reads reports of how U.S. trade treaties have led to depressed wages and substantial job loss in the United States, then the researcher has that knowledge base to use in studying U.S. families dealing with declining wages and lost jobs. The researcher has standing to explore possible shared obliviousness in families concerning the possible involvement of U.S. trade laws in their low family wages or loss of jobs. However, if the researcher is oblivious to the same matters families are oblivious to, the researcher will write an oblivious research report. And there are certainly reasons to think that many social researchers have little interest and capacity in moving beyond the obliviousness of the society in which they live (Chomsky, 2000, pp. 19–20. For example, researchers are influenced by the same forces for obliviousness as anyone else, and they work in a system where grants, publishing, and jobs may be linked to working within the bounds of societal obliviousness.) Then a challenge for a researcher interested in exploring family obliviousness is to establish a knowledge base independent of the families studied and not only to be a good methodologist in studying the families but also to be a good evaluator of the trustworthiness of the independent knowledge base and of the researcher's own understandings of what that knowledge base has to offer.

People's Testimony about Obliviousness in Their Own Family

Looking on the World Wide Web for accounts of family obliviousness, one finds instances of people writing about their own family's obliviousness to a problem habit in the family, a family member being abused, racism in the family, the family's poverty, the looming danger of Nazism in the larger society, or the depression of a family member. Can such accounts of past obliviousness be trusted? When people apply the descriptor "oblivious" to their family, do they mean what is meant in this book by the term? Or might they think that "oblivious" means "downplaying" or "not talking about" the matter? Moreover, just as any outside observer can find it challenging to decide whether a family is oblivious about something, so might a family member who tries to characterize her or his family. And it would be especially challenging to characterize the family as it was some years ago, because memory is selective and also fades and because one may not, in the past, have paid attention to one's family members in ways that would have allowed for a clear sense of whether or not obliviousness to some matter was present. Perhaps family members were aware of the matter but did not talk about it. Perhaps the person testifying about her or his family does not remember accurately or was oblivious to the ways the family was not oblivious. There is evidence that family members may perceive more unanimity than there actually is (Cashmore & Goodnow, 1985), so perhaps even if most members of the family were oblivious some were not. Or perhaps it is defensiveness or vengefulness or the desire to say something definitive and extreme that leads to someone characterizing her or his family as oblivious, when the family was not in fact oblivious.

From another angle, however, one of the most promising ways to study family obliviousness may be to learn from families about their experiences of moving out of obliviousness. One might research a family's accounts of how certain life experiences moved them out of obliviousness (for example, how family therapy, an economic disaster, or the new realizations, learning, and discoveries by one family member eventually ended obliviousness in the entire family). One might research family accounts of the discomforts and joys of leaving some area of obliviousness behind. One might research how a family understands the obliviousness they once had, what it was like to be oblivious, where they think the obliviousness came from and how it was maintained, and how it was a family experience. One might research how leaving one area of obliviousness behind leads to other changes in a family, for example, becoming more interested (or less interested) in ending other areas of obliviousness, changing the community in which they participate, trying to convert others from obliviousness, or revising their sense of their family history.

The Structure of Shared Family Obliviousness

So far in this book I have written as though areas of obliviousness exist in parallel, as though a family could be oblivious to its own privileges and also oblivious to ethically questionable things its government does, to a family member's drug addiction, and so on, with each area of obliviousness independent of the others. But there may be a structure to a family's obliviousness such that the areas of obliviousness are linked, with some areas foundational to others or defended in part by (or because of) their links to others.

Let's say, for example, that for many white families in the United States obliviousness to skin privilege is foundational to obliviousness about what the U.S. government has done to harm people of color in ways that advantage white people. That is, some white families work at being oblivious to government actions that have harmed people of color and advantaged white families, because that could lead them to see themselves as privileged. Similarly, the family's obliviousness to skin privilege may also be guarded by the family's obliviousness to the opinion of some that they are to some extent responsible for the pain, disadvantage, and poverty experienced by families of color. That is, if an oblivious family were to be aware of the accusations that they are in some way responsible for the pain, disadvantage, and poverty of others, that would threaten their obliviousness to their privilege. The structure of these areas of obliviousness may take another shape than what has been suggested here, and perhaps other areas of obliviousness than those named would be crucial to the structure for some families. But the point is that some areas of shared family obliviousness may by foundational to or linked to others.

If there is a structure to family obliviousness, then a challenge in researching an area of obliviousness is that without information about that structure a researcher can be blocked, misled, or confused. First of all, even if a researcher finds a way to learn about a particular area of family obliviousness, there may be no evidence and hence no understanding of the place of that area in the structure of the family's obliviousness. Secondly, the structure of obliviousness could make it much more difficult to learn about any specific area of obliviousness. The ways that an area of obliviousness is guarded in family functioning may seem formidable, and the depth of motivation for guarding it may seem intense. And it may be puzzling to a researcher that so much is invested in protecting what might seem to be an area of shared family obliviousness of no special importance. However, if a researcher manages, by accident or by research skill, to enter into an exploration of a foundational area of family obliviousness, the researcher may encounter great hostility from the family or one form or another of withdrawal from the research (for example, no longer answering questions fully and honestly or refusing to go forward in the research).

The structure of obliviousness may also link areas of obliviousness to areas of knowledge. It is not only that attention to one thing reduces the chances that, because people can only attend to so much, there would be attention to something else. It is that attention to some specific matters directly blocks or makes more or less impossible attention to other specific matters. For example, a family's considerable attention, time, and money put into attending to nutritional health may be structurally linked to the family's efforts to remain oblivious to environmental toxins and pollution. Thinking about nutrition provides all the explanations the family needs for good health or ill health, and hence the family is blocked from attending to air quality, water quality, and toxic materials in the soil.

Obliviousness in Systems in Which the Researcher Is a Member

Then there is the matter of the obliviousness of the systems in which the observer functions. How does a person in an oblivious system research obliviousness? Researchers have their own social systems, and these social systems are inevitably oblivious to a great deal (Chomsky, 2000, pp. 19–20). That is, the systems of researchers (disciplinary systems, university systems, academic department systems, research collaboration systems, and research grant systems) would have their own areas of shared obliviousness. The researcher's own family system obliviousness might also be an impediment in researching obliviousness in other families. Moreover, if the researcher lives in the same society as the people researched, the researcher will be pushed to be oblivious by the same societal forces that push the people researched. For example, people researching family income may be oblivious to illegal income sources (burglary, drug dealing, bribery, embezzlement) to the extent that there are forces in the society (for example, banks that do money laundering, businesses that benefit from bribes) that push the entire society toward obliviousness to illegal income. Even the assumptive system a society imposes on everyone will push researchers toward obliviousness. People in a society in which monogamy is presumed and who are studying marriage may be oblivious to plural marriage involving people they are studying; and people in a society where it is assumed that all children are with their legal parents or guardians and who are studying child custody may be oblivious to the informal movement of children from household to household. Nor are researchers free of the obliviousness and defenses of obliviousness that protect people in families and in the larger society from threats to their privilege, status, comfort, reputation, and peace of mind. If a researcher is embedded in a racial, ethnic, social class, gender, family, sexual orientation, etc. group that says some people are more interesting and more important than others, it would not be surprising if the researcher's studies reflected values and realities of that peer group. Even scholars who claim to be lovingly working for the well-being of others may be oblivious to how their methods and focus deny the realities, experiences, self-respect, and perspectives of those they claim to lovingly work to benefit (Ortega, 2006).

Research systems generally value obliviousness in part because obliviousness to presumably irrelevant or difficult-to-assess issues allows focus on what researchers deem important. And within a research project, obliviousness helps to keep the research manageable in terms of the amount of data to be analyzed. A research group that accumulates too much information about too many matters could be swamped by the surfeit of information and find it very difficult to decide what to make of the data. For example, survey researchers who gather answers to specific questions might be closed to respondent comments on and criticisms of those questions and respondent narratives that move outside the framework of the questions. Many social researchers will try to obtain comparable data from all individuals and families in a study, which means that they work at being oblivious to what is unique to particular individuals and families. One can see that ignoring uniqueness facilitates research; one can also see that it denies the humanity of others by making them simplified, objectified cases (Townley, 2006). And by losing track of that humanity researchers may lose a great deal of important knowledge about the area of research interest.

The standards a researcher may be constrained to follow (by disciplinary colleagues, granting agencies, journal editorial boards, tenure and promotion processes) may demand obliviousness. For example, the enforcement of standards of scientific objectivity by making the researcher appear to be a neutral observer limits how much a social researcher may challenge obliviousness underlying research standards. And the demand to respect the scholarly canon, while in many ways laudable, means that carrying out social research in any area in which the canon has staked a place almost demands an acceptance of much if not all of the obliviousness of that canon. Thus, we cannot usually count on social scientific research to illuminate areas of disciplinary obliviousness. On the contrary, we can usually count on most social scientific research to maintain disciplinary obliviousness; that is, researchers typically work within a disciplinary status quo that makes it difficult to research or to even consider issues outside of that status quo (Argyris, 1980; Harding, 2006; Kuhn, 1970). However, as Kuhn (1970) argued, revolutions can arise in scientific thinking such that areas that had been matters of obliviousness open up to investigation. For example, in the family field, feminist thought has challenged disciplinary obliviousness to women's realities and experiences, opening up new theoretical avenues, and that has stimulated considerable research about gender in families (Osmond & Thorne, 1993).

From another angle, in the past half-century much has occurred in the climate surrounding the social sciences in the United States that makes a focus on self-interested ignorance a source of discomfort for many scholars (Harding, 2006). One reason for this is that, although Karl Marx and Sigmund Freud exist as crucial scholarly sources on self-interested ignorance, McCarthyism and its aftermath frightened many social scientists into distancing themselves from the work of Marx and Freud, with a consequence that the Marxian and Freudian ideas of self-interested ignorance have often been ignored in the social sciences (Harding, 2006). Also, although the work of Marx and Freud on self-interested ignorance has had a substantial impact on the work of social critics, scholars in cultural studies, certain feminist scholars, and others, many scholars have adopted a selective ignorance that supports obliviousness to the power of the unconscious and of bourgeois interests in shaping what their research focuses on, how they carry out their research, and what interpretative practices and outcomes make sense to them (Harding, 2006). Part of the problem is that in recent decades, motivated in part by fear of McCarthyite or similar attacks, the social sciences have to some extent lost their autonomy (Harding, 2006). One consequence of that is arguably that social scientists often selectively ignore the extent to which Marxian and neo-Marxian perspectives on class, gender, race, imperialism, etc. shape and limit scholarly standpoints (Harding, 2006), so that not only is our work cut out for us in trying to understand and deal with family obliviousness, we are also restricted in our ability to understand and deal with our own disciplinary and societal bases of obliviousness in carrying out our research.

Why Researchers Might Gain from Attending to Shared Family Obliviousness

This chapter and the entire book offer an extended argument for the value of researchers' attention to family obliviousness. I would like to add that because so much of family research assumes that family members can and will tell what is going on, what has happened, what they think, what they observe in themselves, and the like, it is a bit of a paradigm shift to move toward studying family obliviousness, matters that family members cannot self-report. But if we do not explore what families are oblivious to, we will ourselves be oblivious to much that is of importance in families.

Feminist, qualitative, and other researchers have fought vigorously to attend to and establish the value of attending to the voices of ordinary people, rather than allowing the expert social scientist voice to dominate research reports. Would a move to focus on family obliviousness weaken support for the voices of family members in research reports? I hope not. It seems consistent with respecting and valuing the diverse voices of family members to study obliviousness in ways that allow family members to speak of their awarenesses, and perhaps their obliviousnesses. Let's pay attention when family members say, "I don't know," "I never thought about that," "We never talked about that," "We are clueless." And then, depending on the research question, it might be appropriate to explore what has been an area of obliviousness. "Why do you think your father never talked about his childhood?" "Do you have any thoughts about what would turn up if he had talked about it?" "Are there ways you could explore his childhood without relying on him for information, and, if so, why do you think you might not have?"

9

Shared Obliviousness That Is Not Quite Shared or Oblivious

Family obliviousness that appears to be totally shared and totally oblivious will at times be less than total. Sometimes a family is oblivious to a matter that in the past was known to at least one family member. Sometimes families erect facades of obliviousness, colluding in an apparent obliviousness that hides that there is at least partial knowledge by some or all family members. In some families there are nonoblivious back channels so that someone in the family can monitor what the family is ostensibly oblivious to. Some families have information specialists who are knowledgeable about matters that nobody else in the family knows. So a family can appear to be oblivious when in fact at least one person in the family was not or is not oblivious, and that imperfect obliviousness may benefit the family.

The Struggle to Achieve or Restore Family Obliviousness

Family members may disagree about what is worth attending to and what is not, so some members may be aware of what others are oblivious to. Family members may disagree about what something means, and in the diversity of their meanings some may have perspectives, information, and awarenesses that others lack. There is no research literature to tell us how often family conflict and disagreement are in part or entirely about whether the family should be oblivious to something, but I am sure that such conflict and disagreement occurs. For example, following the death of a family member, the surviving family members may disagree about the cause of death. One family member, but not others, may be aware that there is a substantial research literature on, say, medical racism in the United States (cf. Rosenblatt & Wallace, 2005), which could have led to the death. The other family members may eventually silence the family member who thinks the death might be due to medical racism. In the end, all family members may seem oblivious to information about medical racism. So after months of family wrangling about the cause of the family member's death, the survivors may all agree that he lived a good life and died of old age, and the family members who had thought that his death might have been due to medical racism may now agree with everyone else about the cause of death. The interpersonal disagreements (overt or not) about the cause of death may eventually be resolved by what appears to be family-wide obliviousness.

It is possible that family members have one reality when together (shared obliviousness) and another when separate (Nadeau, 1998). But it is also possible that the combination of family dynamics (including social pressures, conflict, and the use of power) and individual dispositions (the desire to get along or the valuing of family unity, for example) will lead to shared obliviousness when family members are separate as well as when they are together.

Sometimes what may occur when there is imperfect obliviousness in a family is not about differences in family members' beliefs but about some, though not all, family members' exposure to information that challenges family obliviousness. Then the family may struggle to achieve obliviousness or to return to obliviousness. For example, imagine that family members think of the political party they all support as by far the most moral party in the nation, but it seems that almost every week a leader of their party is named in the news as someone who has committed immoral or illegal acts. For family members who follow the news, each time there is a news report of another leader of the party being caught, accused, indicted, tried, or convicted, there may be a new struggle to go back to being oblivious to the widespread immorality of their party's leaders. One can imagine various ways individual family members struggle to restore their obliviousness or to head off future attacks to their obliviousness (for example, reading only certain publications or deciding that the news sources are extremely biased), but at times some family members will not be oblivious to the upsetting news. It may be that family members do not work in unison but that some take the lead, because they are the ones who receive more of the obliviousness-challenging news or are more eager to hold on to their beliefs about the party. Or perhaps they are more concerned about other family members being influenced by the news, or would be more threatened by a shift by the whole family out of the current family reality. Then it would be they who pressured other family members to read only certain sources, to think about the news media in certain ways, or to forget what they have learned about the failings of their political favorites. On the other hand, many or all family members may value unity and feel uncomfortable with disagreement, so they change themselves or try to change others in order to achieve, maintain, or restore what seems to them to be unity.

More generally, there may be information that challenges a family's constructed realities. Once that information has been accessed, traces of it or even well-organized and very substantial bodies of information related to that alterative reality remain potentially available even if they do not currently constitute what family members think about (Rosenblatt & Wright, 1984). Information that is accessible but not currently in awareness has the potential to become the basis of a shift to a very different family reality, unless individual family members and the family as a whole work at distancing, staying away from, or ignoring the information. Alternative realities can be a very substantial threat. But the alternative realities may also be not so much threatening as simply recognizable and familiar as a different way of thinking about things. For example, a family's current reality may be something such as, "We get along well, and are happy together." But they may have enough information to have an alternative and even comfortably familiar reality such as, "We don't get along particularly well and are emotionally distant from each other much of the time." Perhaps some families would have to be oblivious to one of the apparently inconsistent realities in order to function, but others may not see the two realities as inconsistent and may function well while entertaining both of the realities.

It is not necessarily a simple matter for a family to wipe things out of family discourse and the awareness of individual family members. Some or all family members may remember what they used to think. Some family members may critically examine family changes or current family thinking. New information or reminders may emerge in unpredictable and more or less uncontrollable ways that stimulate new thinking. And to achieve the inattention and unawareness of obliviousness might require a certain amount of critical attention and awareness to oneself or perhaps even more importantly to other family members. For example, a white family may need to police obliviousness to skin privilege and the benefits that come with that. The policing may involve chiding or teasing a family member who reads something from a black perspective. It may involve family-wide coercive opposition to a family member's interracial dating. It may involve squelching a family member's curiosity about a black, Native American, Latino, or Asian American perspective on the family system and the larger systems in which it functions. Family members doing the policing may be oblivious to the perspective that they are trying to block others from learning about, but they must have some awareness of what it is they are supposedly oblivious to in order to recognize that someone is deviating from family obliviousness norms. Of course it is also true that if socialization for obliviousness to skin privilege has been effective, people often will police themselves (Karis, 2006). For example, a white person will stop herself or himself from becoming curious about black experience.

Power, Privilege, and Image Maintenance via Obliviousness

The underlying force for family obliviousness may involve a certain amount of awareness by at least one person in the family that obliviousness can protect individual or family comfort, power, privilege, or image. It is possible that the obliviousness can be achieved without conscious awareness of such matters, but my guess is that some of the time family obliviousness in the service of comfort, power, privilege, or image was a desired state for someone who was clearly aware of the stakes. I can imagine, for example, grandparents who realized that in order to maintain their religious valuing of charity and to also maintain their image of themselves as charitable it would be best to be oblivious to the real economic needs of the poor in their community, including those who had been blocked by racism from economic advancement. The grandparents wanted to think of themselves as charitable but did not want to disperse most or all of their economic resources to help the needy. So the grandparents might have pushed for family obliviousness to the needs of such people because if the family were fully aware of those needs they would think it desirable to donate every cent they had to those people. But perhaps a focus here on the grandparents is too kind to the rest of the family. Perhaps every adult family member in every generation of a well-off family has recognized that to acknowledge the poverty and neediness of others in the community, and the injustice that holds some people in poverty, might call for draining the family's wealth in an effort to help those in need. And rather than give up their wealth, these members of the family, and the family functioning collectively, choose to be oblivious to most or all of the needs of most or all of the poor and of injustices that keep many people poor.

Collusive Obliviousness

One clue that indicates that sometimes family obliviousness is not quite as total and complete as it seems is that when a family is oblivious to matters that might be difficult for them, they seem to be working together to maintain their obliviousness. For example, everyone in the family of someone who has committed a notorious crime may work at being oblivious to information about the crime or to memories that might make it seem that the family member who committed it was actually guilty and in fact had demonstrated the potential, long before the crime was committed, to commit such a crime. So the family might not only work at being oblivious to the horrid side of the family member who committed the crime but also to their own failure to try to block the wrongdoer's path before great harm was done.

I suppose it is possible to collude without awareness, but I think family collusion often involves enough coordination that it requires a degree of awareness. For example, even if Grandmother is the family member who most wants the family to act as though the crime never happened, others will have to have some awareness of what they say to family members and even what they think to be able to go along with what Grandmother wants. So then there are obliviousness facades, where everyone in the family seems genuinely oblivious but in fact they know more than they let on. Related to this, I believe that awareness can be present at one time (as when a family member first learns that he or she is not supposed to notice that Dad is often drunk) and not at others (after they have all learned well that they are not supposed to notice Dad's drunkenness, the not-noticing becomes automatic and not something to think about).

Sampling the "Headlines"

There is a place between obliviousness and awareness in which I think many families operate with regard to some topics. It involves family members sampling or skimming succinct, overview information on certain matters, but not going beyond that succinct, overview information. Just as newspaper readers may read headlines but not articles, television viewers read the program listings or quickly skim through all the channels, or book buyers look at the titles in a bookstore without stopping to open most of the books, family members may skim the "headlines" of their world. They may glance at what is going on around them, but rarely pause to go deeper to find out more. They may Google a topic and scroll through hundreds of hits without selecting any URL for closer examination. They may sort through today's mail and discard most of it without opening it. They may have conversations with one another that skim personal "headlines" for the day, but do not go deeper than that. They may ask their children how the day was in school, receive a one-minute report of "headlines," and be satisfied.

This kind of sampling or skimming may be an important compromise between (1) obliviousness to everything other than what is vitally important and (2) trying to process far more information than is possible to process. By sampling or skimming as many as a thousand times more information sources than can be paid close attention to in a day, family members can be better prepared to deal with things that should be dealt with but that are outside of the usual run of things to pay attention to. They can more easily catch on when a family member is in some kind of trouble or something important to the family has come up. They can receive a degree of reassurance that the context in which they function has not changed in ways that call for a change in what needs close attention.

Sampling or skimming the "headlines" probably means that most of the time family members share obliviousness to most of the details in the stories under the headlines. Also, family members can be in error in assuming they know what a specific "headline" actually means. "Headlines" are simplified, selective, possibly distorted, and open to misunderstanding. So headline sampling may maintain an illusion of awareness among family members but in reality family obliviousness rules guarantee that family members will never go beyond almost any headline and actually may lack accurate information from the headlines they have skimmed. Thus, headline sampling may extend the potential for overcoming certain areas of obliviousness or it can be part of how obliviousness is maintained.

The metaphor of skimming the headlines can also be applied to how families deal with emotionally charged information about family members. There are families in which obliviousness to a specific emotionally charged matter is desired, for example, Grandmother's victimization during the Holocaust or Grandfather's role as an SS officer killing others during the Holocaust. Younger family members may know the headlines of the stories about a grandparent, but obliviousness rules prevent them from trying to know more (even when they seem to be trying to know more) or from actually knowing more. But with their awareness of the headlines, they may fill in the gaps in their knowledge with intensely painful fantasies (Rosenthal, 1998; Rosenthal & Volter, 1998). Their fantasies may coincide with what a grandparent actually experienced, which suggests that even though a family works hard at maintaining obliviousness there are ways in which the walls of obliviousness can be breached (Rosenthal, 1999; Rosenthal & Volter, 1998). Alternatively, the fantasies might be far from accurate, in which case obliviousness has been maintained, but at a cost to the younger people (Rosenthal & Volter, 1998), for example, their cost in painful feelings or emotional distance from their grandparents.

Nonoblivious Back Channels

I believe that there are, in families, paradoxical back channels that preserve valued system obliviousness while monitoring certain areas in which the entire family is apparently oblivious and values that obliviousness. The back channels are paradoxical in that nonobliviousness may be necessary for obliviousness to continue. So while the family is oblivious it also in some fashion monitors some matters for which obliviousness is desired, so there is not total, completely shared obliviousness. For example, the members of a white family may seem to be oblivious to their white skin privilege, but at some level they may monitor possible threats to that privilege. Thus, they may feel concern when the local supermarket starts demanding two picture ID cards from white people who pay by check, not just from people of color, or when a family of color looks at a house for sale in their previously all-white neighborhood. So while the family is seemingly oblivious to their skin privilege they also monitor possible threats to that privilege, and that monitoring has value to them in terms of family system goals because it enables the family to defend its privilege. Perhaps one person in the family is the keeper of the privilege and does the back channel monitoring, but perhaps everyone in the family has a role in maintaining it. (See also the discussion in chapter 3 about obliviousness to victims.)

From another perspective, family systems are and must be morphogenic (Rosenblatt, 1994, pp. 136–39). By this I mean that they adapt and change, because external conditions change, because at times their functioning is not sufficiently adaptive, and because they have to deal with inevitable changes in the family as people age, become ill, become better educated, and so on. Morphogenesis is ongoing for family systems; family systems do not stop changing. Ideally, morphogenesis is facilitated by obliviousness, in the sense that monitoring what is irrelevant to system goals and functioning distracts from monitoring those matters that are truly important to the morphogenesis that ideally should go on. But there are always questions about getting along in a changing environment, about correcting maladaptive mistakes in the choice of what information to ignore, and about making use of new ways of information processing that may enable the processing of more information than was possible previously. That is, the system must have some way of evaluating its changes and nonchanges. So the ordinary morphogenesis that is part of family system functioning may require some sort of back channel review and judgment about at least some of what the system has been oblivious to and how it processes the information. The back channel review and judgment might be initiated by a family member who cares more about the details, who cares more about meeting certain values, or who is more in touch with feelings of personal or familial dissatisfaction.

I believe that as a family changes, its information, perceptions, and evaluations concerning what is in the back channels change. Furthermore, whatever it is attending to in the back channels changes (because the world changes, the family's needs change, the information sources change). So to add to the paradox of obliviousness that at some level is attentive to what the family is oblivious to, there may often be ways for information about, perceptions of, and evaluations of whatever is in the back channels to change. If, for example, a suburban family is ostensibly oblivious to the toxicity of the chemicals that they and their neighbors apply to lawns in order to keep the lawns dark green and weed free, there still could be an ongoing process of learning about and changing opinions about those chemicals. For years the ongoing process might never be reflected in family conversation, but it might appear, seemingly out of nowhere, when a young grandchild visits. At that point, someone in the family may say, "Let's keep our grandchild indoors today, and let's keep the windows closed." And perhaps without even discussing why that should be done, the others in the family agree. Somehow they have learned that the chemicals are dangerous and might be particularly so to a young child. Or perhaps it suffices to say the additional sentence, "The ChemLawn guy was here this morning." Thus, the family that seems to be oblivious to the toxicity of lawn chemicals has ways to monitor information about lawn chemicals and to act on new information, perceptions, and evaluations when it is relevant to what they care about.

But just as there are attentional economics underlying obliviousness (Schneider, 2006), there must be attentional economics to the back channels. If a family is monitoring areas of obliviousness, only some things can be monitored. A family cannot monitor an infinity of information, but there is an infinity available. How is the selective monitoring organized? One possibility is that there may be a sense that what is more important to monitor in the obliviousness zone are things in the near environment or things that seem more likely to have a big impact on the health, safety, economic well-being, and status of the family and its members. That is based on the assumption that these matters are most important to families. But conceivably there are ecologies in which other matters count more and are the ones that are monitored. And even within a category, there must be selective monitoring. There cannot be monitoring, for example, of all matters that are relevant to health, because there is more out there than can possibly be attended to.

Whatever the system of back channel monitoring of matters to which a family is ostensibly oblivious, it is undoubtedly applied at a relatively low level in the sense that it occupies few of the attentional resources of the entire system. One way to do that at the family level is for the back channel monitoring to be done by one or a small number of family members, perhaps by those family members whose other attentional work is not seen as vital to the ongoing business of the family. These may be people who are not what most family members have been socialized to be. That way, relatively few vital system resources may go into monitoring the back channels. I am thinking of people such as the heretic, the acting-out teenager, the person who has more formal education than others in the family have or want, the person who is more curious than others think is desirable, or the religious deviant. They may be seen as potential or actual troublemakers and may be marginalized because they are deviant culturally and in the family, but they also may serve an important service for the family and their deviant label is a way to keep what they do from drawing approval and interest from others in the family.

Thus, the person who violates family obliviousness rules through awareness of this or that may be serving the family through that awareness. The person may be chided or punished for her or his deviance, but at the same time be so useful to the system that if she or he stopped breaking the obliviousness rules quite possibly someone else in the system would have to take on the role. For example, the member of a family who breaks a family obliviousness rule by speaking up about how Uncle Jack seems excessively interested in being alone with little girls may receive considerable disapproval but may help the family to be adaptively aware of Jack's potential as a threat to little girls and an embarrassment to the family. And if that vigilant family member were not monitoring the possibility of someone's being a sexual predator, another family member would have to do the monitoring.

Obliviousness and Information Specialization

In many human systems there is a division of labor in terms of knowing, learning, and remembering. We live in an age of division of labor in regard to who has what information, not an age in which everyone in a system has the same degree of high-level knowledge (Schneider, 2006). There are information specialists in corporations, universities, health care facilities, restaurant kitchens, law firms, accounting firms, governments, car repair facilities, and even in families. For example, one family member may know far more than anyone else in the family about filling out tax returns; another may be the family expert on health issues or gardening; another may be the one family member who knows the neighbors well. Information specialization reduces family system obliviousness, even though many in the system may be oblivious to what each specialist attends to.

A family system with a great deal of information specialization may be one in which most family members are deliberately oblivious to a great deal: "Why should I bother knowing about cars when my brother the mechanic knows everything there is to know about them?" On the other hand, the expertise of a family information specialist may at times be so seductive that it will attract other family members toward that specialization. For example, a family specialist in cooking may so intrigue other family members that they make efforts to learn more about cooking. This is liable to occur particularly among children in the family, who might want to grow up to be like the information specialist and who might be welcomed more readily than other adults as apprentices and emulators.

Family information specialization, for many reasons, may be self-perpetuating. A person's special information is a base of power and personal identity. Then, too, it may be difficult to help others to join one as a specialist in some area, because to become a specialist may require an enormous amount of learning. And then, from another angle, for anyone to become a special- ist in information about X may require that the individual become oblivi- ous to Y and Z, more than if the individual were a generalist. Because of this, once family members develop a variety of specializations, the division of labor is self-perpetuating in that each specialist becomes, in a sense, depen- dent on the other specialists. The health specialist in the family needs the car mechanic and vice versa.

Living in an era of information specialty, a time when there is great dif- ferentiation among personnel in business, higher education, and other areas of knowledge and expertise, we may be more generally oblivious than people were in the past. Whereas in the past an educated person might know, say, 0.00000001 percent of all there was to know, now there may be nobody who knows more than 0.0000000000000000001 percent. This is partly a matter of an increase in the amount of available knowledge, but it is also the product of how systems organize, whereby any system greater than a certain size (or in an environment in which nonobliviousness in important areas requires a great deal of learning) splintering into information specialists. But it might also reflect a society in which obliviousness is treasured. That is, specializa- tion may occur partly because systems value keeping most people oblivious to most things. So when families divide up functions such that one person is a specialist about X and another about Y but most people are oblivious to both X and Y, the family is reflecting processes in the larger society.

Many kinds of specialization are only a benefit to others if there can be communication between specialist and nonspecialist (Schneider, 2006). For example, physicians in many kinds of specialties have to know how to elicit information from patients, and patients with many kinds of problems have to know how to communicate with physicians. Otherwise, the physician's spe- cialized knowledge may be of little use. Similarly, in families some kinds of specialization require communication between the specialist and others who are largely oblivious. Yes, Mom is the tax specialist in the family, but Dad has to be able to answer Mom's questions about his income sources and charitable contributions.

Information specialization in larger organizations seems to be, gener- ally, a good thing, but in organizations as small as a nuclear family, infor- mation specialization carries certain risks. A country might have thousands of specialists in, say, rural economic development; a corporation that needs mechanical engineers may have many on the payroll and can easily find and hire others if those employees retire; but what if only one person in the fam- ily (and, in fact, the world) is a specialist on, say, the family tree, Grandma's medications, or the family's income tax records? When a knowledgeable fam- ily member dies, that family member's specialized information may be dif- ficult or impossible to replace. The family may be able to find sellers of certain kinds of information, tax expertise, for example. And somebody else in the family may have the time and desire to take over certain kinds of specialization, such as becoming an expert on the neighbors. But some information may be lost forever—family history or information about the family genealogy. So in families where specialization occurs it may be helpful, or even necessary, to train a specialist's replacement before the specialist dies. Some other member of the family, preferably one likely to live for many more years, has to learn the family history or who those people are in the old photo album.

If a family information specialist has nobody in the family with whom to communicate about her or his area of specialization, it not only makes the specialization unhelpful to the family, it isolates the specialist. Sadness and frustration are created from the knowledge that others do not care about what one has worked so hard to know. It can also create considerable interpersonal tension when the specialist needs another to understand something to which the others have been oblivious. For example, if the tax specialist in the family cannot explain to anyone else that they are in trouble because others keep poor financial records, that can be frustrating to the specialist and costly to the family.

Knowing More Than Can Be Said

I believe that there can be times when the members of a family know there is more (about an event, a family problem, the family, the community, the world) than can be said. It may be that they feel that something is missing—information, awareness—though they do not know what it is and cannot even put their sense of what it might be into words. It may be that they know that there are gaps in what they know, though they cannot say more than that they know there are those gaps. In a situation like that, they can be said to be both oblivious and not quite oblivious. For example, they may collectively know there is a deficiency in their relationships with one another as they sit around the Thanksgiving table. But they cannot put into words (in part because they are almost completely oblivious) that what is missing is, let us say, warmth and connection that is more than superficial, or genuine thanks giving for the blessings of life.

The members of a family may also know collectively that although their living the good life as it is defined in their culture and community is not satisfying, they cannot put their dissatisfaction into words. For example, it may be that in the aridity of middle-class living and of obsessive consumer busyness and acquisition, they know that they have hungers that are not satisfied, even if they do not know what those hungers are. They are not totally oblivious that something important is out of awareness, but they may be oblivious to what that something is. Thus, they may be oblivious to the fact that it is their vision of the good life as encompassing busyness and acquisition that is the source of their dissatisfaction. And they may be oblivious to unsatisfied hungers for, say, alternative sources of life meaning, or for more intimate relationships.

The members of a family may watch the news on television every night and feel that there is more than can be said—perhaps about what the news does not cover or how it covers what it covers. Again, they may not be able to put to words their sense of what it is that is wrong with the picture, but in their knowledge that something is missing they are not completely oblivious.

There are those who would say that therapy is often or even always about situations where knowing and words fail us (Frosh, 2004). This perspective challenges the postmodern idea that words and language create our realities and argues for realities that operate even if there are no words for them. From the point of view of a focus on shared family obliviousness, perhaps the lack of words keeps family members oblivious, but even without words they can know that something they cannot put to words is real and important. That would be congruent with the discussion in chapter 2 on language and family obliviousness that suggests that family members may avoid learning words for what they are oblivious to when their lack of vocabulary to talk about those things helps to maintain an obliviousness they somehow desire.

Perhaps there is a definitional issue here, revolving around the question of whether one can say that people are oblivious if they know that they are oblivious to some matter but cannot say what it is. One can argue both sides on the definitional issue. Knowing there is more to know about, say, the emptiness and contradictions of one's family life may mean one knows a lot. But knowing that there is more to know without being able to say more than that may also mean that whatever is a matter of obliviousness is well masked and far from personal and family awareness and attention. In fact, one of a family's defenses of an area of obliviousness may be to allow family members to know enough to know that they do not know something but to block them from moving forward from there toward greater knowing. "We know there is something on the other side of this wall, but we don't know what it is, and there is no way around, through, over, or under the wall." And yet, with a small change in the obliviousness of just one family member or a small amount of help from an outsider to overcome their obliviousness, the family could get to the other side of the wall.

Records

Families may maintain records the contents or meaning of which may be domains of obliviousness, but the records may be a way out of the obliviousness should someone or the family as a whole want it. For example, a family may have an attic full of financial records that detail how the family acquired the land on which it lives and the wealth that sustains it. Perhaps there was once a time when it caused discomfort to know those things—perhaps there was something not moral or just about the land and wealth acquisition—so the family slipped into obliviousness about those matters. But the maintenance of the records could allow an opening into the past that might be desirable at some point—when, for example, a family member seeks a squaring of accounts or honesty about the family's history.

From another angle, the creation or maintenance of family obliviousness may require rooting out and destroying records—photos, love letters, income tax returns, checkbooks, clippings, e-mail directories, and so on. Destroying evidence that might challenge an area of obliviousness seems a powerful strategy, and it is illuminating that such things are done. But it is also illuminating that people who seem to be invested in family obliviousness about some matter keep records that threaten the obliviousness. One hears of German families who learn of a Nazi past from old photos of relatives who were Nazis; one hears of white families in the United States who felt proudly and confidently white but who have learned from old photos that some of their ancestors were black or American Indian. The "memory" trace was quite possibly not in anyone's mind, but it existed in stored material that could be accessed and made sense of. I do not want to be simplistic about stored records. I imagine sometimes records are stored and continue to be kept because nobody knows they exist or what they are, and some people save everything, tattered clothing and refrigerator magnets that advertise events long in the past as well as meaningful photos and old correspondence. But it is possible that some of the time, the saving is in service of someone's sense that someday it may be necessary or desirable to overcome family obliviousness about a matter.

The Future of Shared Family Obliviousness

FAMILIES WILL ALWAYS BE oblivious. They cannot escape from shared family obliviousness. There is vastly more to know than can possibly be known. A family will always ignore aspects of relationships within the family and virtually infinite amounts of information about what goes on outside the family, including some information that would be disruptive or uncomfortable to know. And families will never be on their own in achieving obliviousness. The mainstream media, government officials, corporations, the schools, and other sources of information have a very substantial influence on the patterns of family obliviousness. Moreover, vast quantities of new knowledge are constantly being created and discovered, and this new knowledge makes demands that call for new kinds of family awareness and also new additions to family obliviousness. The Internet, for example, has changed the world for many families in ways that require new kinds of awareness but also new kinds of obliviousness in order to accommodate the new kinds of awareness.

The Future for Families

Shared Family Obliviousness to Aspects of the Future

Many families plan for aspects of the future. They may save for a child's education, plan for retirement, plan a family vacation, and budget for a significant purchase. Many couples plan to have children and use birth control to govern when they will have children and how many they have. Weddings are planned. Insurance policies protect against possibly difficult futures. Education is often seen as an investment in the future. Trips to the supermarket and the food in the refrigerator are testimony to family anticipation of the near future. So families obviously approach some aspects of the future with awareness.

However, each substantial change in the family's situation brings new futures to attend to. Families may be generally aware of the implications for the future of such proximal changes as a pregnancy, a birth, the development of a chronic illness, a divorce, an employment change, a change in a law that substantially affects the family, a child leaving home, or a child marrying. But then there are future changes in community, society, and planet that may be hinted at or even discussed at length in sources of information to which families may be oblivious, rendering them oblivious to the implications of these possible futures for them. The members of a family may not think about how their government will change if energy becomes vastly more expensive or the oceans rise eighty feet, and how those changes might affect their future as a family, or what preparations would be best for that possible future. However, in defense of shared family obliviousness, there are infinitely more possible futures than any family can anticipate. So families have to be oblivious to many possible futures.

To some extent, families are pushed into paying attention to certain areas of the future by decisions made in government, by industries that aggressively market future-related products (such as insurance, retirement accounts, and safety devices), and by how educators frame the future for students. For example, members of middle-class families may have employers who insist that they elect health insurance options, and middle-class families with children in school may be told repeatedly that they have to anticipate and prepare for the educational future of those children. On the other hand, society (as represented by the mainstream media, government leaders, corporate advertising and public relations, and educational institutions) pushes families to ignore many other possible futures. Thus, many families in the United States have had years of preparation for purchasing their next automobile. But U.S. families seem to receive very little encouragement to pay attention to what increasing levels of carcinogenic chemicals in the air and hormones in drinking water may do to their health. However, to the extent that every advance in technology, change in the local and planetary physical environment, change in government, change in education, change in health care, change in family law, and so on changes the future for a family, anticipating the future requires aiming at a moving, blurry, ever-changing, and largely unknowable target. Still, at least some families might be much better prepared for certain conceivable futures if there were not so much control of attention and obliviousness by the institutions of society.

But then families have their own defenses. Some families may react to the latest changes by working at being oblivious to them, or at least trying to have as little awareness of them as possible. Imagine a family living in a coastal area, only a few feet above sea level, tuning out warnings in reports of global warming that predict more frequent severe hurricanes and a substantial rise in the sea level.

Googling the phrase "oblivious family" brings up accounts of families that were oblivious to the future. The family did not know the Holocaust was coming, war was coming, an economic depression was coming. Typically, writers of these World Wide Web accounts of family obliviousness seem critical, even contemptuous of their families. Families are faulted for obliviousness to what they should have known was coming. One can look at the faulting as hindsight bias (e.g., Guilbault, Bryant, Brockway, & Posavac, 2004), which means that to some extent after an outcome is known people think it was more likely to have occurred than it actually was. If hindsight bias has occurred, one could say that the obliviousness was at least as much on the part of the person who sees others to have been inappropriately oblivious as it was on part of the people who are criticized for their obliviousness. One can also look at the faulting as an assertion that because humans have a capacity to predict aspects of the near future with some accuracy, if people were not oblivious they would have known enough to protect themselves and their loved ones from what seemed to be in the offing. Yes, in some instances there may have been enough information available to a criticized family to know they were in great danger. The warfare was coming closer every day; businesses were going bankrupt or closing at a record pace; the Nazis were doing progressively more dreadful things to Jews. In those instances one could say obliviousness was harmful, that confronted with strong evidence, responsible family members seem to have discounted or even ignored the evidence and thereby put themselves and the entire family at risk. It follows from this that the question for a family nowadays is whether there is evidence enough to put aside obliviousness to an approaching disaster and to make changes to protect themselves before it is too late. Furthermore, since many of the writers on the World Wide Web who criticize families for their obliviousness to the future are the children of the people they criticize, one could say that our obliviousness now may not be excused in the future by those who depend on us to protect and guide them.

Expanding Family Information Capacity

This book asserts that because there is a limited information-carrying capacity for any family, and because there are vast amounts of information, families will always have to be oblivious to much. But what if it is not that simple?

What if many families have more information-holding and processing capacity than they use? What if families can learn to handle more information? What if certain life circumstances can push families into a higher level of awareness and a lower level of obliviousness about many matters? What if, in a crisis, information priorities can change radically so that any family can process substantial amounts of information relevant to the crisis. If any of this is so, then a family might be able to gain much more knowledge and awareness without being overloaded.

The reduction of obliviousness does not necessarily require great learning. Recognizing that there is a blind spot in what one knows is a substantial first step in reducing obliviousness (Schneider, 2006). So one way that families might, in the future, systematically extend their capacity to resist obliviousness is to work more effectively at knowing what they do not know. To be able to say, for example, "We as a family do not know a lot about our unverbalized rules that govern what goes on in our family and do not know how we got locked into thinking that we needed lots of material goods," would be an important step in reducing obliviousness, even if the family members only knew that they did not know very much in these areas.

One can imagine, too, that if educational institutions and the mainstream media were to lay out for people the idea of learning what it is one's family is oblivious to, the future for some families might be one of greater capacity to overcome obliviousness. This potential for resistance to obliviousness might be valued by members of those families even if most of the time they only stopped at the point of saying, "We know that in this area there is a lot we don't know." Perhaps realizing that something is not known will lead to new family system rules about how to select what to explore and what to leave alone in areas of known ignorance. In businesses, the criterion that may be most referred to in deciding what areas of ignorance to explore is generally profitability (Schneider, 2006). For families, there could well be noneconomic criteria such as safety, health, family happiness, or family members getting along well. Whatever the criteria, there might always be systematic family resistance to information that could be overloading (cf. Schneider, 2006, writing about business management), but then there also could be means of accessing knowledge in condensed, headline, or other limited forms (Schneider, 2006) that would not be so overloading.

The threats that many families might prefer to deal with by inattention may be increasing in number and scope. For example, advances in science lead to both new discoveries of health hazards and new creation of chemicals and devices that are health hazards. It seems, too, that every day some source in the mass media reports new actions or new plans by the U.S. government that from some perspective are potentially harmful to many families, the environment, and the country. Then there are the threats of global warming, economic decline and instability, increasing (or increasingly obvious) govern-ment corruption, and many other kinds of frightening and upsetting news. To the extent that family obliviousness is partly about avoiding knowledge of what is distressing and threatening, it would not be surprising if the members of many families avoided attending to science news, news about actions of the U.S. government, or economic news, or had ways to misunderstand or quickly forget such news. What irony that with great potential for awareness and knowledge comes great skill at not acquiring awareness and knowledge and quite possibly great skill at dismissing or ignoring new information!

On the other hand, the expansion of family information capacity may not be a concern for many families whose contact with the media is mainly or exclusively with mainstream media. First of all, if most mainstream news media rely mainly on a few dozen news sources (such as the Associated Press, the *New York Times*, *60 Minutes*, the *Washington Post*, and *Slate*), families may not be able to acquire much additional information by going to additional mainstream media sources. What's more, if mainstream television is largely focused on entertainment and advertising, family members may spend many hours each day viewing television programs with little or no chance of acquir-ing new information about what is going on in the world that is a danger, a threat, a benefit, or a source of insight for the family. And their capacity to assimilate new and important information may be limited and not stimu-lated to grow. Similarly, to the extent that newspapers have declined in recent years, with fewer newspapers, fewer independent newspapers, fewer report-ers, and fewer resources devoted to investigative reporting, families who rely on newspapers for information may not hit their capacity for handling sig-nificant family-relevant news or may not be stimulated by newspaper sources to develop their information-carrying capacity.

But then the mainstream press cumulatively offers many examples of good investigative reporting and analysis, both from the high-visibility sources such as the Associated Press, *60 Minutes*, the *New York Times*, and the *Wall Street Journal* and from small market newspapers like the *Dayton Daily News* or the *Toledo Blade*. So even families who attend primarily or exclusively to the mainstream media might gain from paying close attention to those media. They might also gain from an increase in investigative report-ing by these sources. And perhaps just as much, they would gain from these sources getting better at keeping investigative stories alive through aggressive follow-up work. That is, shared obliviousness that is seemingly not good for families might be harder to achieve if there were more follow-up reporting and analysis of important stories. It might be good for families to be helped not to forget, for example, that the Food and Drug Administration is put-ting fewer resources into food and drug inspection and into testing of pro-posed new drugs for harmful side effects, or that much of the Food and Drug

Administration budget for evaluating new drugs comes from the drug companies whose drugs are being evaluated.

All this is not to say that the mainstream press is the only or the most important place for families to acquire new knowledge and to be stimulated to expand their capacity to handle knowledge. Among many potential information sources, the alternative print, television, and electronic media provide considerable news that is covered little if at all in the mainstream media.

The Future of Family Obliviousness and the Future of Families

I have looked at enough books and articles entitled "The Future of" something to believe that many of us who write about the future are oblivious to how much of what we think is in the future is what we know about the present. Some writings anticipate substantial transformations from the present, but many do not. They seem, instead, to project the present into the future. Perhaps that is not so bad, because the future has its roots in the present. However, I hope that what I write here is based on awareness of how unknowable the future is and is open in useful ways to futures I cannot anticipate or imagine.

Just as the environment for families may change what families can or will choose to be oblivious about, what is in the future of families may change the future of obliviousness. Imagine that in the future more households in the United States will include very elderly people, multiple nuclear families, or people who have agreed to be family members together for a certain contractual period. Any of these could change shared family obliviousness about the arbitrariness of what constitutes a family, about who should share a dwelling, about the costs and strangeness of valuing independent individual living, or about the limitations inherent in thinking that family relationships should be lifelong.

Imagine that more families in the United States will have people who are dealing with the familial and personal effects of catastrophic environmental, economic, or political events. That might make it more difficult for families to share obliviousness to the long-term emotional and relationship consequences of loss and trauma.

Imagine that the era of cheap hydrocarbon energy and of family obliviousness in the United States to what burning hydrocarbons does to the planetary ecology will soon be over. So it will no longer be desirable, or perhaps even possible, for most people to travel considerable distance to and from work five or more days a week or for relatives who are dispersed over considerable distances to get together often. Single-family dwellings and low population density commuter suburbs could become uneconomical in terms of energy costs. More people will work at home or very close to home most of the time. Presumably families will not so easily disperse to considerable distances. Moving a thousand miles away to go to college or to live in a more tolerable climate while leaving close relatives behind may mean that one will rarely see those people. Presumably the end of cheap hydrocarbon energy will mean that multistory, multifamily dwellings, not separated by side yards or substantial backyards and integrated with shopping, workplaces, health care, and schools will become much more common. These changes would change family obliviousness about the costs of what they have been considering to be a good life. Family members living nearer to each other than families now-adays do would not necessarily change their obliviousness to what goes on in their family, but living in close proximity to other family members may increase the power of families to police and shape their own shared oblivi-ousness. On the other hand, family members who do move far away might well find it more likely that they will be oblivious to a great deal in the lives of those left behind and will find their own lives more matters of oblivious-ness for those left behind than they might like. Electronic communication may have an impact on reducing family obliviousness, and developments in electronic technologies may make communication from a distance, for many families, easier and more vivid than it is now. But electronic communication has limits that mean that in some ways family members will find it easy to be oblivious to aspects of one another's experiences and feelings.

The obliviousness that goes with skin privilege for white families in the United States may resist change, but there are forces at work that may change that. There is the possibility of popularization of the growing body of literature on white privilege and white obliviousness to that privilege. Fur-thermore, there are ever more occasions when people of color have a voice in the mainstream mass media, education, corporations, and government and are able to speak their truths. Possibly, white obliviousness to white privilege will continue indefinitely. Possibly, it will go into hiding in ways that keep it alive but protect it better from information that white privilege involves unearned advantage and dominance and rests on injustice, ignorance, and harm caused to people of color. But perhaps white obliviousness and white skin privilege will begin to disappear. I think if that happens, we might find that other forms of privilege, and the obliviousness that accompanies them, will increase. I say that because I think that rank and differential privilege will still be important to many U.S. families. So if it is not skin privilege that certain families can rely on and be oblivious about, it might well be some other kind of unearned privilege (and the obliviousness that goes with it) that various families will rely on. On the other hand, recognizing and overcoming obliviousness to skin privilege may make it harder for people to be oblivious to other unearned advantages. If that is so, the decline of obliviousness to skin privilege might lead to a general decline in obliviousness to unearned gender, class, heterosexual, etc. privilege.

Perhaps in the future of families there will be as much tension as there is now between (1) family unity, agreement, likeness of values and purposes, shared images of the family, and shared obliviousness and (2) differentiation of family experience, values, understandings, beliefs, images, purposes, and obliviousness. The tension will, I think, be driven by countervailing forces that are unlikely to go away. The forces for differentiation in the family include the individualistic values of the United States, the diversity of daily experiences of family members, the forces for women's and children's rights, and the many situations in which there is no need for family members to agree because family members are acting as individuals. Also, many families are likely to value and need diversity of family knowledge in various problem-solving situations. The forces for family unanimity will continue and in some ways may become more powerful. I think there will continue to be ideological forces that value family unanimity, and in some circles, a patriarchy in which all go along with the patriarchal views. A key reason why I think the forces for family unanimity will continue is that I think there will be very demanding new situations, for example, weather or economic disasters, in which families will be more likely to stay together if family members share certain kinds of obliviousness. On the other hand, the forces against patriarchal or any other submersion of family diversity may be strong in difficult new situations. In those situations families may benefit from independent thinking. That is, in situations that call for family creativity, flexibility, and problem solving, lockstep response or control by one person might often be maladaptive.

Earlier chapters of this book have made the case for seeing obliviousness as a defense against guilt feelings or other dark thoughts about unearned privilege and about things done by individual family members or the family collectively that have harmed others. I think there is some chance that in the future there will be even more reasons for many families in the United States to feel guilty and otherwise to think dark thoughts about their privilege, actions, and inactions. Imagine, for example, heat waves, famines, and droughts that kill hundreds of millions of people around the world while sparing most who live in the United States. And this at a time when many around the world say that climate changes would not have happened or would not have been so extreme had people in the United States responded effectively to the first warnings of global climate change. In such a situation, perhaps many U.S. families will not tolerate certain kinds of awarenesses.

Undoubtedly there will be enormous changes of all sorts in the future that cannot be anticipated now. Try to imagine, for example, new diseases and radically new approaches to the treatment of some kinds of disease, new chemicals to which to be addicted, new ways of providing education, new occupations, new forms of financial transaction, and new and powerful political forces in the United States and the world. Imagine, too, matters for family attention and/or obliviousness for which there are currently no words in the English language.

A question raised from the beginning of this book is that even if a family system must be oblivious about an enormous amount, why the areas of obliviousness the system has, and why not others? The answer to the question is simple and not simple. What is simple is that one would assume that most families monitor and are aware of what is highest priority to them for comfort, survival, safety, and continued existence. One would think that families must be aware, for example, of how to obtain adequate amounts of nourishing food and safe drinking water. Since attending to anything may ordinarily move at least some other things toward obliviousness, one reason families are oblivious to what they are oblivious to is that they are inclined to tune out a great deal of what is not necessary for comfort, survival, safety, and continued existence.

What is not simple in answering the question about why these areas of obliviousness and not others is that, first of all, obliviousness and attention are to a substantial extent the products of actions by other systems with which a family is in contact. Things the mainstream mass media ignore, that corporate and government leaders do not want people to know, and that the schools and the family's religious congregation never deal with are relatively easy for families to tune out. External systems so drive obliviousness in many families that families may often seem oblivious to information that one would think would be of crucial importance to their well-being and continued existence. The power of the external systems may be reflected in the ways that families might be oblivious to the changes being brought by global warming, the decline in the ozone layer, the increase in toxic chemicals in the air and drinking water, the rising ocean levels, the decline in the world's fisheries and so on. (One does not have to agree about the significance or even the existence of what I have enumerated. One can enumerate other items that many families probably tune out that from some perspectives one would think would be in the best interests of families to attend to.)

On the other hand, there is likely to be variation and change of family obliviousness over time. There are also likely to be family information specialists, family members who deviate from the family obliviousness rules, and various back-channel mechanisms that to some extent monitor what is in an area of obliviousness. So potentially the domains of family obliviousness may change in adaptive, moral, or otherwise desirable ways from time to time. Some families may even delight in finding areas of information to which they had been oblivious. They may, for example, learn about a possible cure for a family member's seemingly incurable illness. They may discover something new about their family history that endows them with an increased sense of pride or a new understanding of how the family became what it is. They may overcome obliviousness to a family member's chemical dependency, go through treatment, and come through to the other side as a much more connecting, loving, and honest family. Even a family coming to understand its own processes of becoming oblivious and of maintaining obliviousness may find a great deal of joy as they are freed from something its members are glad to be freed from—for example, the coercion of a family myth that bad things happen in the family when family members get to know people of other religions or cultures. Some families may even come to understand collectively some of the system dynamics that underlie family obliviousness, and even if that does not end their obliviousness to much, it may emancipate them from certain forces for obliviousness. It may also emancipate them from feeling shame, guilt, and other forms of distress at how much they are or may be oblivious to, because in understanding the dynamics of obliviousness they will also understand the necessity of obliviousness.

From another angle, quite possibly every action, perception, measurement, memory, and judgment has an error rate. From that perspective, obliviousness to much can be understood as a way to cut down on error rates for the actions, perceptions, and so on in which the family engages. By paying attention to less, it is possible for the family to make fewer mistakes. On the other hand, in the near future families may have to learn much that is new. There may be much in the offing that could bring vast changes for families, not the least of it being substantial climate change. Great changes in the environment demand great changes in awareness and obliviousness. Maintaining low error rates by being obliviousness to much may not be such an adaptive strategy in the new and changing environment, which may mean that error rates will go up. Family members will more often miscommunicate, make mistakes that cause accidents, make financial mistakes, and act on the basis of misunderstood information because they will live in a situation that demands that they deal with a great deal. However, to the extent that errors are disheartening and make it harder to act in the future, perhaps one kind of obliviousness that may increase for families in that demanding new environment is obliviousness to errors. That is, families may not be as aware as they would be nowadays of their miscommunications, accidents, financial mistakes, and so on. In a sense, the need to act may become more important than the need to be aware of and correct certain mistakes.

The future of family obliviousness is about the future of every bit of what goes on in a family. Shared obliviousness to this or that affects everything by limiting and shaping what family members do with each other and apart from each other, what they say and how they say it, what they think, what they learn and do not learn, what they focus on and what they ignore, what they devote their energies to, what upsets them and what gratifies them, and how they spend their time. Shared obliviousness is and will be a window into all of family life.

Shared Family Obliviousness as a Topic in College Courses

If one thinks of education as being about pushing into areas of ignorance and lack of awareness, one would hope that much of higher education is about overcoming obliviousness. So students taking courses about families should ideally be learning to see their own families and the families of others in new, insightful ways. They should, ideally, become more aware of family dynamics, the context and history of families, the nonobvious ways that families get into trouble, the diversity of families, and everything else that is important in understanding families. So it seems necessary to deal explicitly with shared obliviousness in families. In fact, there are college-level family courses that deal with matters addressed in this book, for example, the obliviousness of families with skin, class, and other kinds of privilege. Adding new discussions about the dynamics of shared family obliviousness and the necessity of that shared obliviousness may be helpful to students in understanding their own families and in learning to work with the families of others. The question of why a family is oblivious to X and not Y can open up interesting, stimulating, important lines of thinking. However, it can also create great trouble for a student and the student's family. Sometimes trouble is good, inasmuch as the trouble and the process of dealing with it get people to places about which they feel very good. Sometimes trouble is just trouble.

When I teach about obliviousness I work at not persuading students to change or to change their families. For example, when I teach about white obliviousness to skin privilege, I make it clear that I am not trying to persuade any student to think or act differently or to go home and challenge family members. But I think it is my obligation, given the courses I teach, to teach about shared family obliviousness as it relates to privilege, and I tell students that even if I do not need to persuade them to change, I do want them to understand the content I teach.

I imagine that the ethical and pedagogical position I take allows white students to shrug off the content. However, I also know that some white students take the content to heart personally in that it stimulates them to reflect in new ways about themselves and their world. A few talk about bringing the content home to their families, and some of them say that their family members "get" the content. But I imagine that for other white students, all the forces that guard family obliviousness could come into play.

From some students of color, I hear an appreciation and understanding of matters they have known and that family members also know. But from others I hear about a different kind of shared obliviousness, that they and the members of their families were not aware of how entrenched white obliviousness to skin privilege may be.

I think it is appropriate to teach students who take courses dealing with families about family obliviousness in many different content areas. Why wouldn't it be good for students to understand that there are people who want their family to be oblivious to the forces that lead to high health care costs? Why wouldn't it be good for students to understand government policies concerning lending, credit, debt, bankruptcy, and financial institution freedom to hide information and mislead borrowers and investors?

However, I am very cautious about what might come of teaching students to probe obliviousness to matters within their own families. Probing family secrets and myths, shared family obliviousness about a problem habit or the ways family members avoid intimacy may be wonderfully freeing. But it may also set off the family equivalent of a nuclear bomb, which could destroy the family or create serious fallout problems and years of struggle with scar tissue. My inclination is to cover those matters in ways that do not push students to reflect on their own family experience. One of the important lessons to learn about shared family obliviousness is that it is well defended, so to some extent I do not think I need to worry about students breaching a wall of family obliviousness or a student who breaches a wall getting very far. But I do not think I am paid to push students and their families into onerous situations or to change any student's family.

The Future of Theorizing about Family Obliviousness

I think of this book as a work of theory; every chapter offers theoretical assertions and speculations about shared family obliviousness. I think an important part of the future of theorizing about shared family obliviousness is to explore the issues raised in this book and as with any conceptual area, to challenge, to push toward greater clarity, to test implications, to add qualifications and complexities, and to explore alternative theoretical framings. And I believe an important part of the theoretical work to be to further explicate the theoretical structure, assumptions, implications, and so on that go with the various kinds of obliviousness this book discusses.

Although the emphasis in this book is on shared family obliviousness, the book offers ideas about different patterns of obliviousness in families, including a number of patterns in which obliviousness is not shared fully by all family members. Theorizing needs to account for the diversity of patterns and for the processes of moving from one pattern to another. To advance the discussion of patterning of family obliviousness, what follows is a partial list of types of shared obliviousness that could be generated from discussions in this book. This is not the only list that could be generated from this book, and one can even see from how some of the categories in the list are described that this partial list could be used to generate additional patterns. Future research and theory may explore these patterns and the processes of moving from one to another, of maintaining shared obliviousness, and of moving into shared obliviousness.

1. All family members have been totally oblivious all along.
2. Some family member has not been oblivious or totally oblivious, whereas others have been.
3. The adults are knowledgeable (for example, about a shared secret), but the children are not (or vice versa, or more generally, some subgroup shares a secret that others do not).
4. Family members have inklings, but these do not rise to the level of knowledge or awareness.
5. Family members are aware of something important but not of its meanings.
6. Family members are aware of an important matter and of its meanings but not of the ethical implications of the matter.
7. Someone in the family once knew something, or maybe even several did, but now it seems to be forgotten or no longer thought about.
8. The family is oblivious about some aspect of their family of origin's past, which means in the past family members (who are dead, perhaps) were aware.
9. Family members think they know something, but what they know is a myth.
10. Family members think they know something, but they are wrong.
11. Family members would know something if they paid attention to it, because the evidence is present for them to know, but they never pay attention to it.
12. Major areas of a family's obliviousness are shared widely in their community or culture (as opposed to major areas of a family's obliviousness being unique to the one family).
13. The family is oblivious that they are oblivious about a lot or to something specific (as opposed to being aware that they are oblivious to a lot or to something specific).
14. The family's shared obliviousness arises from dynamics of relationships within the family (as opposed to dynamics in the individual members separately or dynamics in the society and community).
15. Shared family obliviousness to some matter is motivated (e.g., to avoid feelings of guilt, fear, or shame) versus the shared family obliviousness seems to have no motivational foundation.
16. Shared areas of shared family obliviousness are normative in family development (e.g., parent obliviousness about aspects of the child's life and vice versa).

17. The family shares obliviousness to what is very difficult for anyone to know.
18. Shared family obliviousness is a failure of curiosity (versus a failure of information sources).
19. Shared family obliviousness helps a family in some ways (versus shared family obliviousness makes trouble for a family in some ways, versus shared family obliviousness neither helps nor makes trouble, versus shared family obliviousness both helps and makes trouble).
20. Shared family obliviousness hurts others outside of the family (versus shared family obliviousness does not hurt others outside of the family).

How Can Families Help Themselves?

Of course families can help themselves. They do it all the time. I am sure that many/most/perhaps all families have helped themselves in some ways through their shared obliviousness and through their movement in and out of shared obliviousness. But then I think it possible that many families have been in trouble because of their shared family obliviousness or their struggles about it. Can a family turn to a book like this for help? I wish they would not. This is not a self-help or family help book. I don't know enough about family obliviousness, let alone about any specific family, to offer confidently helpful ideas. I think family obliviousness is a complex matter, and self-help or family help books often seem to me to be dangerous because they ignore complexity. But then, I also believe that people can sometimes find help in the oddest places. I imagine that some people can find something helpful in almost any chapter of this book. Also, there are a few core ideas in this book that I imagine would speak to some who are looking for help with an issue of family obliviousness.

Simplify

Every family is oblivious to a vast amount of information, and shared family obliviousness is good in the sense that it is necessary for families not to be overloaded with information. On the other hand, there are no doubt families who are overwhelmed but who have no choice. They must, for now, deal with too much. Still, even some of those families, plus perhaps all other families who seem to be overwhelmed, might be better off if they could simplify and push more of the matters they have been dealing with into zones of oblivious. They may benefit from simplifying what needs to be paid attention to, investigated, decided, done, communicated to others, remembered, managed, and discussed within the family. That means they gain if they can stop paying attention to so much, stop learning so much, stop keeping track of so much information, stop caring about so much. They can gain if they simplify record keeping, bring fewer news sources into the family, shorten their to-do lists, and get rid of more of their paper mail and e-mail without even paying attention to it. There may be an increase in feelings of overload as the family makes choices about how to simplify and what to be oblivious to, but once the simplification has occurred, things may be easier.

Specialize

Families who are overloaded by information might benefit from specialization, a division of labor about who attends to what and who works at being oblivious to what. For the overloaded family, whether the topic is keeping track of the income tax, watching the weather report each night, or attending all the meetings of an organization together, there may be a desired togetherness in jointly processing all that information. But if the family is overloaded with information, it might benefit from one family member becoming the one who keeps on top of income tax, another one who watches the weather forecast and who can offer a quick summary of it to the others, and another one who attends a specific organization meeting and who can briefly summarize what went on for the others.

Families who are in trouble because they share obliviousness to matters it would be best to be aware of might benefit from differentiating what they are oblivious to. If the family members collectively have the will and the capacity to put aside some areas of shared obliviousness but not the time, they could develop specializations. What if, for example, nobody has been paying attention to addiction issues even though Grandma seems to be in serious trouble with an addiction to a prescription medication? What if, for example, nobody has been attentive to how a child in the family has been letting homework slide? If there is too much there for everyone to monitor, one possibility is to agree that one adult family member will pay attention to Grandma's addiction issues and another will pay attention to the child's homework issues.

Accept Obliviousness in Self, Others, and the Family

Understanding the forces for shared family obliviousness and the necessity of it, people can understand how and why there is resistance (in self, other family members, and the entire family) to their becoming less oblivious than they currently are. They can understand, and perhaps accept, that in their roles in the family they are forces for obliviousness for themselves and other family members. Once people understand the family system roots of certain kinds of shared obliviousness they may be in a better place to contend against some of the undesirable consequences of those kinds of shared obliviousness, to minimize the costs of that shared obliviousness, or to increase their capacity to move out of obliviousness when by some standards that would be desirable. But they may also be in a much better place to accept that there are limits to what can be attended to and known, that family systems are always oblivious to vast amounts of information and must be, and that there are great costs to pay for trying to attend to too much.

References

Aarts, P. G. H. (1998). Intergenerational effects in families of World War II survivors from the Dutch East Indies: Aftermath of another Dutch war. In Y. Danieli (Ed.), *International handbook of multigenerational legacies of trauma* (pp. 175–87). New York: Plenum.

Abrams, M. S. (1999). Intergenerational transmission of trauma: Recent contributions from the literature of family systems approaches to treatment. *American Journal of Psychotherapy, 53*, 225–31.

Alexander, A. (2001). The meaning of television in the American family. In J. Bryant & J. A. Bryant (Eds.), *Television and the American family*, 2nd ed. (pp. 273–87). Mahwah, NJ: LEA.

Alford, C. F. (2001). *Whistleblowers: Broken lives and organizational power*. Ithaca: Cornell University Press.

Alford, C. F. (2007). Whistle-blower narratives: The experience of choiceless choice. *Social Research, 74*, 223–48.

Ancharoff, M. R., Munroe, J. F., & Fisher, L. M. (1998). The legacy of combat trauma: Clinical implications of intergenerational transmission. In Y. Danieli (Ed.), *International handbook of multigenerational legacies of trauma* (pp. 257–76). New York: Plenum.

Anderson, C. M., & Stewart, S. (1983). *Mastering resistance: A practical guide to family therapy*. New York: Guilford.

Apraku, K. K. (1996). *Outside looking in: An African perspective on American pluralistic society*. Westport: Praeger.

Argyris, C. (1980). *Inner contradictions of rigorous research*. New York: Academic Press.

Arnow, P. (2007). From self-censorship to official censorship. *Extra! 20*(2), 9–10.

Baker, K. G., & Gippenreiter, J. B. (1998). Stalin's purge and its impact on Russian families: A pilot study. In Y. Danieli (Ed.), *International handbook of multigenerational legacies of trauma* (pp. 403–34). New York: Plenum.

Baker, R. (2002). What are they hiding? *The Nation, 274*(7), 11–16.

Barner, J. R., & Rosenblatt, P. C. (2008) Giving at a loss: Couple exchange after the death of a parent. *Mortality, 13*, 318–34.

Bateson, G. (1972). *Steps to an ecology of mind*. New York: Ballantine.

Bateson, G. (1980). *Mind and nature: A necessary unity*. New York: Bantam.

Bausch, K. C. (2001). *The emerging consensus in social systems theory*. New York: Kluwer Academic.

Beattie, K. (1998). *The scar that binds: American culture and the Vietnam War*. New York: New York University Press.

Becker, E. (1973). *The denial of death*. New York: Free Press.

Bennett, S. E. (2003). Is the public's ignorance of politics trivial? *Critical Review, 15,* 307–37.

Bennett, W. L, Lawrence, R. G., & Livingston, S. (2007). *When the press fails: Political power and the news media from Iraq to Katrina*. Chicago: University of Chicago Press.

Benson, J. E. (1990). Good neighbors: Ethnic relations in Garden City trailer courts. *Urban Anthropology, 19,* 361–86.

Bergantino, L. (1997). Existential family therapy: Personal power-parental authority-effective action-freedom. *Contemporary Family Therapy: An International Journal, 19,* 383–90.

Berger, P. L., & Kellner, H. (1964). Marriage and the construction of reality. *Diogenes, 46,* 1–24.

Berger, P. L., & Luckmann, T. (1966). *The social construction of reality*. New York: Doubleday.

Berkowitz, B. (2007). Lincoln Group. *Z Magazine, 20*(1), 10–13.

Berkowitz, D. A. (1977). On the reclaiming of denied affects in family therapy. *Family Process, 16,* 495–501.

Bernstein, P. P., Duncan, S. W., Gavin, L. A., Lindahl, K. M., & Ozonoff, S. (1989). Resistance to psychotherapy after a child dies: The effects of the death on parents and siblings. *Psychotherapy: Theory, Research, Practice, Training, 26,* 227–32.

Bertalanffy, L. von (1986). *General systems theory: Foundations, development, applications*. New York: Braziller.

Bianculli, D. (2000). *Teleliteracy: Taking television seriously*. Syracuse: Syracuse University Press.

Blair, D., & McNamara, G. (1997). *Ripples on a cosmic sea: The search for gravitational waves*. Reading, MA: Addison-Wesley.

Bloom, M. V. (1980). *Adolescent-parental separation*. New York: Gardner.

Blumenthal, S. (2006). Where torture got him. *Progressive Populist, 12*(18), 13.

Boszormenyi-Nagy, I., & Spark, G. (1973). *Invisible loyalties*. New York: Harper & Row.

Bowen, M. (1978). *Family therapy in clinical practice*. New York: Jason Aronson.

Bowker, G. C. (2005). *Memory practices in the sciences*. Cambridge: MIT Press.

Brooks, G. R. (2001). Developing gender awareness: When therapist growth promotes family growth. In S. H. McDaniel, D.-D. Lusterman, & C. L. Philpot (Eds.),

Casebook for integrating family therapy: An ecosystemic approach (pp. 265–74). Washington, DC: American Psychological Association Press.

Brosio, R. A. (1994). *A radical democratic critique of capitalist education*. New York: Peter Lang.

Brown-Smith, N. (1998). Family secrets. *Journal of Family Issues, 19*, 20–42.

Bucher, R. E. (1985). The stories nobody tells: The family myth in the life of the neurotic [in Portuguese]. *Psicologia: Teoria e Pesquisa, 1*, 7–18.

Buckley, W. (1967). *Sociology and modern systems theory*. Englewood Cliffs, NJ: Prentice-Hall.

Burgoon, J. K., Berger, C. R., & Waldron, V. R. (2000). Mindfulness and interpersonal communication. *Journal of Social Issues, 56*, 105–27.

Burton, B., & Fansetta, D. (2006). The worst media manipulations—and how to win against the spin. In P. Phillips & Project Censored (Eds.), *Censored 2007: The top 25 censored stories* (pp. 267–81). New York: Seven Stories Press.

Byng-Hall, J. (1973). Family myths used as defence in conjoint family therapy. *British Journal of Medical Psychology, 46*, 239–50.

Byng-Hall, J., & Thompson, P. (1990). The power of family myths. In R. Samuel & P. Thompson (Eds.), *The myths we live by* (pp. 216–24). New York: Routledge.

Byrne, B. (2006). In search of a "Good mix": 'Race,' class and practices of mothering. *Sociology, 40*, 1001–17.

Campbell, D. T., & Fiske, D. W. (1959). Convergent and discriminant validation by the multitrait-multimethod matrix. *Psychological Bulletin, 56*, 81–105.

Carey, A. (1997). *Taking the risk out of democracy: Corporate propaganda versus freedom and liberty*. Urbana: University of Illinois Press.

Cashmore, J. A., & Goodnow, J. J. (1985). Agreement between generations: A two process approach. *Child Development, 56*, 493–501.

Cerulo, K. A. (2006). *Never saw it coming: Cultural challenges to envisioning the worst*. Chicago: University of Chicago Press.

Chambers, D. (2001). *Representing the family*. London: Sage.

Chomsky, N. (1989). *Necessary illusions: Thought control in democratic societies*. Boston: South End Press.

Chomsky, N. (1996). *Class warfare: Interviews with David Barsamian*. Monroe, ME: Common Courage Press.

Chomsky, N. (2000). *Chomsky on miseducation*. Lanham, MD: Rowman & Littlefield.

Chomsky, N. (2008). We own the world. *Z Magazine, 21*(1), 25–31.

Christian, D., Pufahl, I., & Rhodes, N. C. (2005). Fostering foreign language proficiency: What the U.S. can learn from other countries. *Phi Delta Kappan, 87*, 226–28.

Cohen, S. (2001). *States of denial: Knowing about atrocities and suffering*. Malden, MA: Blackwell.

Coleman, H. (1992). "Good families don't…" (and other family myths). *Journal of Child and Youth Care, 7*(2), 59–68.

Conley, D. (2001). Universal freckle, or how I learned to be white. In B. B. Rasmussen, E. Kleinberg, I. J. Nexica, & M. Wray (Eds.), *The making and unmaking of whiteness* (pp. 25–42). Durham: Duke University Press.

Cooklin, A., & Gorell Barnes, G. (1993). Taboos and social order: New encounters for family and therapist. In E. Imber-Black (Ed). *Secrets in families and family therapy.* (pp. 292–328). New York: Norton.

Corning, P. A., & Kline, S. J. (1998). Thermodynamics, information, and life revisited, Part II: "Thermoeconomics" and "control information." *Systems Research and Behavioral Science, 15*, 453–82.

Coupland, J. (2000). Introduction: Sociolinguist perspectives on small talk. In J. Coupland (Ed.), *Small talk* (pp. 1–25). Harlow, England: Pearson Education.

Crago, H. (1998). The unconscious of the individual and the unconscious of the system. *Australian and New Zealand Journal of Family Therapy, 19*(2), iii–iv.

Crapanzano, V. (1986). *Waiting: The whites of South Africa.* New York: Vintage.

Dalmage, H. M. (2004). Protecting racial comfort, protecting white privilege. In H. M. Dalmage (Ed.), *The politics of multiracialism: Challenging racial thinking* (pp. 203–18). Albany: State University of New York Press.

Dana, J. (2005). Conflicts of interest and strategic ignorance of harm. In D. A. Moore, D. M. Cain, G. Loewenstein, & M. H. Bazerman (Eds.), *Conflicts of interest: Challenges and solutions in business, law, medicine, and public policy* (pp. 206–23). New York: Cambridge University Press.

Davies, N. J. S. (2007). From aggression to genocide. *Z Magazine, 20*(9), 49–54.

DeFrancisco, V. L. (1991). The sounds of silence: How men silence women in marital relations. *Discourse and Society, 2*, 413–23.

DeMuth, D. H. (1994). A global paradigm for family therapists in the 21st century. In B. B. Gould & D. H. DeMuth (Eds.), *The global family therapist: Integrating the personal, professional, and political* (pp. 3–21). Boston: Allyn & Bacon.

Derdyn, A. P., & Waters, D. B. (1981). Unshared loss and marital conflict. *Journal of Marital and Family Therapy, 7*, 481–87.

DeVree, J. K. (1994). Information in nature, human behavior, and social life. *Behavioral Science, 39*, 117–36.

Doherty, G. (1976). Dying and the conspiracy of denial. *Essence, 1*, 34–37.

Donzelot, J. (1979). *The policing of families.* New York: Pantheon.

Doster, A. (2007). Education reform: Pass or fail? *In These Times, 32*(2), 42–45.

Dumas, J. E. (2005). Mindfulness-based parent training: Strategies to lessen the grip of automaticity in families with disruptive children. *Journal of Clinical Child and Adolescent Psychology, 34*, 779–91.

Ehrenhaus, P. (1993). Cultural narratives and the therapeutic motif: The political containment of Vietnam veterans. In D. K. Mumby (Ed.), *Narrative and social control: Political perspectives* (pp. 77–96). Newbury Park, CA: Sage.

Ehrich, K. R., & Irwin, J. R. (2005). Willful ignorance in the request for product attribute information. *Journal of Marketing Research, 42*, 266–77.

Emery, K., & Ohanian, S. (2004). *Why is corporate America bashing our public schools.* Portsmouth, NH: Heinemann.

Engeler, E. (2006, November 3). UN says 2005 set greenhouse gas record. *Seattle Post-Intelligencer.* Retrieved March 16, 2007, from http://seattlepi.com/national/292839_climate18.html.

Farah, D. (1999). Papers show U.S. role in Guatemalan abuses: In declassified documents, diplomats describe massacres, CIA ties to army. *International Journal of Health Services, 29*, 897–99.

Feldman, L. B. (1980). Marital conflict and marital intimacy: An integrative psychodynamic-behavioral-systemic model. *Advances in Family Psychiatry, 2*, 19–33.

Festinger, L. (1957). *A theory of cognitive dissonance.* Evanston, IL: Row Peterson.

Fisher, R., & Ury, W. (1981). *Getting to yes: Negotiating agreement without giving in.* New York: Houghton Mifflin.

Flaskas, C. (2002). *Family therapy beyond postmodernism.* New York: Routledge.

Flood, R. L. (1999). *Rethinking the fifth discipline: Learning with the unknowable.* New York: Routledge.

Ford, F. R, (1983). Rules: The invisible family. *Family Process, 22*, 135–45.

Framo, J. L. (1981). The integration of marital therapy with sessions with family of origin. In A. S. Gurman & D. P. Kniskern (Eds.), *Handbook of family therapy* (pp. 133–58). New York: Brunner/Mazel.

Foucault, M. (1980). *Power/knowledge: Selected interviews and other writings.* New York: Pantheon.

Fox, J. O. (2001). *If Americans really understood the income tax: Uncovering our most expensive ignorance.* Boulder, CO: Westview.

Fraad, H. (1996/1997). At home with incest. *Rethinking Marxism, 9*(4), 16–39.

Frankenberg, R. (1993). *White women, race matters: The social construction of whiteness.* Minneapolis: University of Minnesota Press.

Frankenberg, R. (2001). The mirage of an unmarked whiteness. In B. B. Rasmussen, E. Kleinberg, I. J. Nexica, & M. Wray (Eds.), *The making and unmaking of whiteness* (pp. 72–96). Durham: Duke University Press.

Freud, S. (1924). *A general introduction to psychoanalysis.* Boston: Boni & Liveright.

Freud, S. (1960). *Group psychology and the analysis of the ego.* New York: Bantam.

Frick, F. C. (1959). Information theory. In S. Koch (Ed.), *Psychology: A study of a science,* vol. 2 (pp. 611–36). New York: McGraw-Hill.

Fromm, E. (1941). *Escape from freedom.* New York: Farrar & Reinhart.

Frosh, S. (2004). Knowing more than we can say. In D. A. Pare & G. Larner (Eds.), *Collaborative practice in psychology and therapy* (pp. 55–68). Binghamton, NY: Haworth.

Gantz, W. (2001). Conflicts and resolution strategies associated with television in marital life. In J. Bryant & J. A. Bryant (Eds.), *Television and the American family*, 2nd ed. (pp. 289–316). Mahwah, NJ: LEA.

Gelbspan, R. (2004). *Boiling point: How politicians, big oil and coal, journalists, and activists are fueling the climate crisis—and what we can do to avert disaster.* New York: Basic Books.

Gerhart, D. R., & McCollum, E. E. (2007). Engaging suffering: Towards a mindful re-visioning of family therapy practice. *Journal of Marital and Family Therapy, 33,* 214–26.

Gilbert, K. R. (1996). "We've had the same loss, why don't we have the same grief?" Loss and differential grief in families. *Death Studies, 20,* 269–83.

Glick, I. D., & Kessler, D. R. (1974). *Marital and family therapy.* New York: Grune & Stratton.

Goffman, E. (1959). *The presentation of self in everyday life.* New York: Doubleday Anchor.

Goffman, E. (1974). *Frame analysis.* New York: Harper & Row.

Goodell, J. (2007, November 1). The prophet of climate change: James Lovelock. *Rolling Stone.* Retrieved April 16, 2008, from http://www.rollingstone.com:80/politics/story/16956300/the_prophet_of_climate_change_james_lovelock

Goodman, A. (2007). Prosecuting Nixon could have elevated the nation. *Progressive Populist, 13*(2), 22.

Gubrium, J. F., & Holstein, J. A. (1990). *What is family?* Mountain View, CA: Mayfield.

Guilbault, R. L., Bryant, F. B., Brockway, J. H., & Posavac, E. J. (2004). A meta-analysis of research on hindsight bias. *Basic and Applied Social Psychology, 26,* 103–17.

Hall, J. (2000). It hurts to be a girl: Growing up poor, white, and female. *Gender and Society, 14,* 630–43.

Hankiss, E. (2006). *The toothpaste of immortality: Self-construction in the consumer age.* Washington, DC: Woodrow Wilson Center Press.

Hanks, R. S. (1993). Rethinking family decision making: A family decision making model under constraints on time and information. *Marriage and Family Review, 18*(3–4), 223–40.

Harding, S. (2006). Two influential theories of ignorance and philosophy's interests in ignoring them. *Hypatia, 21,* 20–36.

Hare-Mustin, R. T. (1994). Discourses in the mirrored room: A postmodern analysis of therapy. *Family Process, 33,* 19–35.

Harootunian, H. (2004). Shadowing history: National narratives and the persistence of the everyday. *Cultural Studies, 18,* 181–200.

Hecker, M. (1993). Family reconstruction in Germany: An attempt to confront the past. In B. Heimannsberg & C. J. Schmidt (Eds.), *Collective silence: German identity and the legacy of shame* (pp. 73–93). San Francisco: Jossey-Bass.

Herman, E. S. (2006). Language and institutional perversions in a time of painful birth pangs. *Z Magazine, 19*(10), 27–31.

Hermida, J.-R. F., Villa, R. S., Seco, G. V, & Perez, J.-M. E. (2003). Evaluation of what parents know about their children's drug use and how they perceive the most common family risk factors. *Journal of Drug Education, 33*, 337–53.

Hill, D. (2005). State theory and the neoliberal reconstruction of schooling and teacher education. In G. E. Fischman, P. McLaren, H. Sunker, & C. Lankshear (Eds.), *Critical theories, radical pedagogies, and global conflicts* (pp. 23–51). Lanham, MD: Rowman & Littlefield.

Hitchens, E. (1972). Denial: An identified theme in marital relationships of sex offenders. *Perspectives in Psychiatric Care, 10*(4), 152–59.

Hobart, M. (1993). Introduction: The growth of ignorance? In M. Hobart (Ed.), *An anthropological critique of development: The growth of ignorance* (pp. 1–30). London: Routledge.

Hoke, S. L., Sykes, C., & Winn, M. (1989). Systemic/strategic interventions targeting denial in the incestuous family. *Journal of Strategic and Systemic Therapies, 8*(4), 44–51.

Hollar, J., Jackson, J., & Goldstein, H. (2006). Fear and favor 2005: FAIR's sixth annual report. In P. Phillips & Project Censored (Eds.), *Censored 2007: The top 25 censored stories* (pp. 283–94). New York: Seven Stories Press.

hooks, b. (1992). Representations of whiteness in the black imagination. In b. hooks, *Black looks: Race and representation* (pp. 165–78). Boston: South End Press.

Hopper, E. (1996). The social unconscious in clinical work. *Group, 20*, 7–42.

Horowitz, G. J. (1991). *In the shadow of death: Living outside the gates of Mauthausen.* London: Tauris.

Horton-Deutsch, S. L., & Horton, J. M. (2003). Mindfulness: Overcoming intractable conflict. *Archives of Psychiatric Nursing, 17*, 186–93.

Imber-Black, E. (1990). Multiple embedded systems. In M. P. Mirkin (Ed.), *The social and political contexts of family therapy* (pp. 3–18). Boston: Allyn & Bacon.

Imber-Black, E. (1993). *Secrets in families and family therapy.* New York: Norton.

Ireland, D. (2006). The tragedy of Gary Webb. *In These Times, 30*(10), 43–44.

Jackson, J. K. (1962). Alcoholism and the family. In D. J. Pittman & C. R. Snyder (Eds.), *Society, culture, and drinking patterns* (pp. 472–92). Carbondale: Southern Illinois University Press.

Jackson, M. C. (1991). Social systems theory and practice: The need for a critical approach. In R. L. Flood & M. C. Jackson (Eds.), *Critical systems thinking: Directed readings* (pp. 117–38). New York: Wiley.

Jensen, R. (2006). Introduction. In P. Phillips & Project Censored (Eds.), *Censored 2007: The top 25 censored stories* (pp. 27–31). New York: Seven Stories Press.

Kantor, D., & Lehr, W. (1975). *Inside the family: Toward a theory of family process.* San Francisco: Jossey-Bass.

Karis, T. A. (2004). "I prefer to speak of culture": White mothers of multiracial children. In H. M. Dalmage (Ed.), *The politics of multiracialism: Challenging racial thinking* (pp. 161–74). Albany: State University of New York Press.

Karis, T. A. (2006). The psychology of whiteness: Moving beyond separation to connection. Unpublished paper presented at the Oxford Round Table on Global Security in the 21st Century, University of Oxford, Oxford, England, August 8, 2006.

Karpel, M. A. (1980). Family secrets: I. Conceptual and ethical issues in the relational context, II. Ethical and practical considerations in therapeutic management. *Family Process, 19*, 295–306.

Katz, F. E. (1993). *Ordinary people and extraordinary evil.* Albany: State University of New York Press.

Killian, K. D. (2002). Dominant and marginalized discourses in interracial couples' narratives: Implications for family therapists. *Family Process, 41*, 603–18.

Kinzer, S. (2006). *Overthrow: America's century of regime change from Hawaii to Iraq.* New York: Times Books, Henry Holt.

Klinger, E. (1977). *Meaning and void: Inner experience and the incentives in people's lives.* Minneapolis: University of Minnesota Press.

Kolhatkar, S., & Ingalls, J. (2006). *Bleeding Afghanistan: Washington, warlords, and the propaganda of silence.* New York: Seven Stories Press.

Kornbluh, P. (1999). Chile declassified: Newly released documents reveal close US ties to the Pinochet regime. *The Nation, 269*(5), 21–24.

Kornbluh, P. (2003). *The Pinochet file: A declassified dossier on atrocity and accountability.* New York: New Press.

Krestan, J., & Bepko, C. (1993). On lies, secrets, and silence: The multiple levels of denial in addictive families. In E. Imber-Black (Ed). *Secrets in families and family therapy.* (pp. 141–59). New York: Norton.

Krondorfer, B. (1994). Our soul has not suffered: Intimacy and hostility between fathers and sons in post-Shoah Germany. *Journal of Men's Studies, 2*, 209–20.

Kuhn, T. S. (1970). *The structure of scientific revolutions*, 2nd ed. Chicago: University of Chicago Press.

Lacey, C., & Longman, D. (1997). *The press as public educator.* Luton, United Kingdom: University of Luton Press.

Lakoff, G., & Johnson, M. (1980*). Metaphors we live by.* Chicago: University of Chicago Press.

Lang, M. (1996). Silence therapy with Holocaust survivors and their families. *Australian and New Zealand Journal of Family Therapy, 16*, 1–10.

Lantz, J. (1992). Resistance in family logotherapy. *Contemporary Family Therapy, 14*, 405–18.

Lawrence, D. B. (1999). *The economic value of information.* New York: Springer.

Liechty, D. (2002). Introduction. In D. Liechty (Ed.), *Death and denial: Interdisciplinary perspectives on the legacy of Ernest Becker* (pp. ix–xvi). Westport, CT: Praeger.

Limbert, W. M., & Bullock, H. E. (2005). "Playing the fool": U.S. welfare policy from a critical race theory perspective. *Feminism and Psychology, 15*, 253–74.

Lipsitz, G. (2006). *The possessive investment in whiteness: How white people profit from identity politics*, rev. ed. Philadelphia: Temple University Press.

Long, E. C. J., Angera, J. J., & Hakoyama, M. (2006). Using videotaped feedback during intervention with married couples: A qualitative assessment. *Family Relations, 55*, 428–38.

Lutz, C. A., & Collins, J. L. (1993). *Reading* National Geographic. Chicago: University of Chicago Press.

Manning, P. (2001). *News and news sources: A critical introduction*. London: Sage.

Margolis, E., Soldatenko, M., Acker, S, & Gair, M. (2001). Hiding and outing the curriculum. In E. Margolis (Ed.), *The hidden curriculum in higher education* (pp. 1–19). New York: Routledge.

Martin, B. (2007). *Justice ignited: The dynamics of backfire*. Lanham, MD: Rowman & Littlefield.

Martinot, S. (2003). *The rule of racialization: Class, identity, governance*. Philadelphia: Temple University Press.

Maturana, H. (1987). Everything is said by an observer. In W. I. Thompson (Ed.), *Gaia, a way of knowing: Political implications of the new biology* (pp. 65–82). Great Barrington, MA: Lindisfarne.

Maturana, H. R., & Varela, F. G. (1987). *The tree of knowledge: The biological roots of human understanding*. Boston: New Science Library.

McCown, W. G., & Johnson, J. (1993). The treatment resistant family. In W. G. McCown, J. Johnson, & Associates (Eds.) *Therapy with treatment resistant families* (pp. 1–21). New York: Haworth.

McCown, W. G., Johnson, J., & Carise, D. (1993). Family problem solving: A crisis intervention/consultation approach. In W. G. McCown, J. Johnson, & Associates (Eds.) *Therapy with treatment resistant families* (pp. 167–200). New York: Haworth

McDonough, S. C. (2000). Interaction guidance: An approach for difficult-to-engage families. In C. H. Zeanah, Jr. (Ed.), *Handbook of infant mental health*, 2nd ed. (pp. 485–93). New York: Guilford.

McIntosh, P. (1988). *White privilege and male privilege: A personal account of coming to see correspondences through work in women's studies*. Working Paper, No. 189. Wellesley College Center for Research on Women, 23 pp.

Metzger, L. (1988). *From denial to recovery: Counseling problem drinkers, alcoholics, and their families*. San Francisco: Jossey-Bass.

Midgley, G. (1996). What is this thing called CST? In R. L. Flood & N. R. A. Romm (Eds.), *Critical systems thinking: Current research and practice* (pp. 7–24). New York: Plenum.

Miller, J. G. (1965). Living systems: Structure and process. *Behavioral Science, 10*, 337–399.

Miller, M. B., Bernstein, H., & Sharkey, H. (1973). Denial of parental illness and maintenance of familial homeostasis. *Journal of the American Geriatrics Society, 21,* 278–85.

Minuchin, S., & Fishman, H. C. (1981). *Family therapy techniques.* Cambridge: Harvard University Press.

Mitchell, P. R., & Schoeffel, J. (Eds.) (2002). *Understanding power: The indispensable Chomsky.* New York: The New Press.

Montgomery, J., & Fewer, W. (1988). *Family systems and beyond.* New York: Human Sciences Press.

Moon, D. (1999). White enculturation and bourgeois ideology: The discursive production of "good (white) girls." In T. K. Nakayama & J. N. Martin, (Eds.), *Whiteness: The communication of social identity* (pp. 177–97). Thousand Oaks, CA: Sage.

Mor, N. (1990). Holocaust messages from the past. *Contemporary Family Therapy: An International Journal, 12,* 371–79.

Nadeau, J. W. (1998). *Families making sense of death.* Beverly Hills, CA: Sage.

Nagata, D. K. (1990). The Japanese American internment: Exploring the transgenerational consequences of traumatic stress. *Journal of Traumatic Stress, 31,* 47–69.

Nagata, D. K. (1991). Transgenerational impact of the Japanese-American internment: Clinical issues in working with children of former internees. *Psychotherapy: Theory, Research, Practice, Training, 28,* 121–28.

Nagata, D. K. (1998). Intergenerational effects of the Japanese American internment. In Y. Danieli (Ed.), *International handbook of multigenerational legacies of trauma* (pp. 125–39). New York: Plenum.

Neisser, U. (1967). *Cognitive psychology.* New York: Appleton-Century-Crofts.

Nevins, J. (2005). *A not-so-distant horror: Mass violence in East Timor.* Ithaca: Cornell University Press.

Nydegger, C. N., & Mitteness, L. S. (1988). Etiquette and ritual in family conversation. *American Behavioral Scientist, 31,* 702–16.

Ortega, M. (2006). Being lovingly, knowingly ignorant: White feminism and women of color. *Hypatia, 21*(3), 56–74

Osmond, M. W., & Thorne, B. (1993). Feminist theories: The social construction of gender in families and society. In P. G. Boss, W. J. Doherty, R. LaRossa, W. R. Schumm, & S. K. Steinmetz (Eds.), *Sourcebook of family theories and methods: A contextual approach* (pp. 591–623). New York: Plenum.

Outlaw, L. T., Jr. (2007). Social ordering and the systematic production of ignorance. In S. Sullivan & N. Tuana (Eds.), *Race and epistemologies of ignorance* (pp. 197–211). Albany: State University of New York Press.

Overton, D. (1994). Why counseling is not sought in deteriorating relationships: The effect of denial. *British Journal of Guidance and Counselling, 22,* 405–16.

Paolucci, B., Hall, O. A., & Axinn, N. (1977). *Family decision making: An ecosystem approach.* New York: Wiley.

Papp, P. (1993). The worm in the bud: Secrets between parents and children. In E. Imber-Black (Ed.), *Secrets in families and family therapy* (pp. 66–85). New York: Norton.

Parenti, M. (1997). Introduction: Methods of media manipulation. In C. Jensen & Project Censored (Eds.), *Twenty years of censored news* (pp. 27–31). New York: Seven Stories Press.

Paul, N. L., & Grosser, G. H. (1991). Operational mourning and its role in conjoint family therapy. In F. Walsh & M. McGoldrick (Eds.), *Living beyond loss: Death in the family* (pp. 93–103). New York: Norton.

Pedelty, M. (1995). *War stories: The culture of foreign correspondents.* New York: Routledge.

Pew Internet and American Life Project (2007). Washington, DC: Pew Internet and American Life Project. Retrieved February 18, 2008 from http://poll.orspub.com/document.php?id=quest07.out_24625&type=hitlist&num=26.

Philo, G. (2004). The mass production of ignorance: News content and audience understanding. In C. Paterson & A. Sreberny (Eds.), *International news in the 21st century* (pp. 199–224). Eastleigh, United Kingdom: John Libbey Publishing for University of Luton Press.

Pinderhughes, E. (1990). Legacy of slavery: The experience of black families in America. In M. P. Mirkin (Ed.), *The social and political contexts of family therapy* (pp. 289–305). Boston: Allyn & Bacon.

Podesta, J. (2003). Bush's secret government: Using fear and national security to hide information from the public. *The American Prospect, 14*(8), 44–46.

Pollner, M., & McDonald-Wikler, L. (1985). The social construction of unreality: A case study of a family's attribution of competence to a severely retarded child. *Family Process, 24,* 241–54.

Pring, G. W., & Canan, P. (1996, March 29). Slapp-happy companies. New York *Times,* p. 21.

Proctor, R. N. (1995). *Cancer wars: How politics shapes what we know and don't know about cancer.* New York: Basic Books.

Quick, D. D. (2006, April 3). SLAPP suits infringe on constitutional rights [Letter to the editor]. *Michigan Lawyers Weekly.* Retrieved October 26, 2006 from Lexis-Nexis Academic.

Raby, R. (2005). Polite, well-dressed, and on time: Secondary school conduct codes and the production of docile citizens. *Canadian Review of Sociology and Anthropology, 42,* 71–91.

Retsinas, J. (2007). A manual for cutting Medicaid. *Progressive Populist, 13*(19), 15.

Reusser, J. W., & Murphy, B. C. (1990). Family therapy in the nuclear age: From clinical to global. In M. P. Mirkin (Ed.), *The social and political contexts of family therapy* (pp. 395–407). Boston: Allyn & Bacon.

Rich, A. (1979). Disloyal to civilization: Feminism, racism, gynophobia. In A. Rich, *On lies, secrets, and silence* (pp. 275–310). New York: Norton

Ridley, B. (1989). Family response in head injury: Denial...or hope for the future? *Social Science and Medicine, 29*, 555–61.

Roberts, G., & Klibanoff, H. (2006). *The race beat: The press, the civil rights struggle, and the awakening of a nation.* New York: Knopf.

Rogal, B. (2003). CHA seals records on relocation. *The Chicago Reporter, 32* (1), 6–7.

Ropohl, A., Elstner, S., Hensen, J., & Harsch, I. A. (2005). A "shrinking" patient: An endocrine disorder? *General Hospital Psychiatry, 27*, 150–52.

Rosenau, P. M. (1992). *Post-modernism and the social sciences.* Princeton: Princeton University Press.

Rosenberg, H. (2004). *Not so prime time: Chasing the trivial of American television.* Chicago: Ivan R. Dee.

Rosenblatt, P. C. (1994). *The metaphors of family systems theory.* New York: Guilford.

Rosenblatt, P. C. (1999). Ethics of qualitative interviewing in grieving families. In A. Memon & R. Bull (Eds.), *Handbook of the psychology of interviewing* (pp. 197–209). Chichester, Sussex, Great Britain: John Wiley & Sons.

Rosenblatt, P. C. (2000). *Parent grief: Narratives of loss and relationship.* New York: Brunner/Routledge.

Rosenblatt, P. C. (2001). A social constructionist perspective on cultural differences in grief. In M. S. Stroebe, R. O. Hansson, W. Stroebe, & H. Schut (Eds.), *Handbook of bereavement research: Consequences, coping, and care* (pp. 285–300). Washington, DC: American Psychological Association Press.

Rosenblatt, P. C. (2003). Bereavement in cross-cultural perspective. In C. D. Bryant (Ed.), *Handbook of death and dying*, vol. 2 (pp. 855–61). Thousand Oaks, CA: Sage.

Rosenblatt, P. C. (2006). *Two in a bed: The social system of couple bed sharing.* Albany: State University of New York Press.

Rosenblatt, P. C., & Burns, L. H. (1986). Long term effects of perinatal loss. *Journal of Family Issues, 7*, 237–53.

Rosenblatt, P. C., & Cunningham, M. R. (1976). Television watching and family tension. *Journal of Marriage and the Family, 38*, 105–11.

Rosenblatt, P. C., Karis, T. A., & Powell, R. D. (1995). *Multiracial couples: Black and white voices.* Thousand Oaks, CA: Sage.

Rosenblatt, P. C., & Wallace, B. R. (2005). *African American grief.* New York: Brunner-Routledge.

Rosenblatt, P. C., & Wright, S. E. (1984). Shadow realities in close relationships. *American Journal of Family Therapy, 12*(2), 45–54.

Rosenthal, G. (Ed.) (1998). *The Holocaust in three generations: Families of victims and perpetrators of the Nazi regime.* London: Cassell.

Rosenthal, G., & Volter, B. (1998). Three generations in Jewish and non-Jewish German families after the unification of Germany. In Y. Danieli (Ed.), *International handbook of multigenerational legacies of trauma* (pp. 297–313). New York: Plenum.

Rothenberg, P. (2000). *Invisible privilege: A memoir about race, class, and gender.* Lawrence: University Press of Kansas.

Rubinowitz, L. S., & Perry, I. (2001/2002). Crimes without punishment: White neighbors' resistance to black entry. *Journal of Criminal Law and Criminology, 92,* 335–428.

Sager, C. J., Kaplan, H. S., Gundlach, R. H., Kremer, M., Lenz, R., & Royce, J. R. (1971). The marriage contract. *Family Process, 10,* 311–26.

Salam, R. (2001). The confounding state: Public ignorance and the politics of identity. *Critical Review, 14,* 299–325.

Sandler, J. (2007). The war on whistle-blowers. *Progressive Populist, 13* (21), 1, 8–9.

Scahill, J. (2007). Bush's shadow army. *The Nation, 284*(13), 11, 13–14, 16, 18–20.

Scharf, D. E., & Scharf, J. S. (2005). Psychodynamic couple therapy. In G. O. Gabbard, J. S. Beck, & J. Holmes (Eds.), *Oxford textbook of psychotherapy* (pp. 67–75). New York: Oxford University Press.

Scharf, D. E., & Scharf, J. S. (2007). Family as the link between individual and social origins of prejudice. In H. Parens, A. Mahfouz, S. W. Twemlow, & D. E. Scharf (Eds.), *The future of prejudice: Psychoanalysis and the prevention of prejudice* (pp. 97–110). Lanham, MD: Rowman & Littlefield.

Scharf, J. S., & Scharf, D. E. (2004). Guest editorial, special issue: Object relations couple and family therapy. *International Journal of Applied Psychoanalytic Studies, 1,* 211–13.

Schindler, H. (2005). A Nazi in the family closet? *Australian and New Zealand Journal of Family Therapy, 26,* 165–68.

Schneider, D. M. (1980). *American kinship: A cultural account,* 2nd ed. Chicago: University of Chicago Press.

Schneider, U. (2006). The other side of the distinction: The management of ignorance. In B. Renzl, K. Matzler, & H. Hinterhuber (Eds.), *The future of knowledge management* (pp. 99–111). New York: Palgrave Macmillan.

Schudson, M. (2000). America's ignorant voters. *Wilson Quarterly, 24,* 16–22.

Seale, C. (1995). Heroic death. *Sociology, 29,* 597–613.

Seattle *Post-Intelligencer* (2007, March 8). U.S. bars talk of climate change effects on bears. Retrieved March 8, 2007 from http://seattlepi.nwsource.com/national/306820_bears09.html.

Shaddock, D. (1998). *From impasse to intimacy: How understanding unconscious needs can transform relationships.* Northvale, NJ: Jason Aronson.

Shamai, M., & Lev, R. (1999). Marital quality among couples living under the threat of forced relocation: The case of families in the Golan Heights. *Journal of Marital and Family Therapy, 25,* 237–52.

Sharps, M. J., & Martin, S. S. (2002). "Mindless" decision making as a failure of contextual reasoning. *Journal of Psychology: Interdisciplinary and Applied, 136,* 272–82.

Sillars, A., & Kalbflesch, P. M. (1989). Implicit and explicit decision-making styles in couples. In D. Brinberg & J. Jaccard (Eds.), *Dyadic decision making* (pp. 179–215). New York: Springer Verlag.

Simon, R. (1990). Does discussion of nuclear war have a place in family therapy? In M. P. Mirkin (Ed.), *The social and political contexts of family therapy* (pp. 383–94). Boston: Allyn & Bacon.

Soley, L. (2002). *Censorship, Inc.: The corporate threat to free speech in the United States.* New York: Monthly Review Press.

Stacks, J. F. (2003/04). Hard times for hard news: A clinical look at U.S. foreign coverage. *World Policy Journal, 20*(4), 12–21.

Stein, H. F. (1985). The unfolding: A clinical tragedy in two acts, or—when life is like a night at the opera. In H. F. Stein & M. Apprey (Eds.), *Context and dynamics in clinical knowledge* (pp. 106–24). Charlottesville: University Press of Virginia.

Steyn, M. (1999). White identity in context: A personal narratives. In T. K. Nakayama & J. N. Martin, (Eds.), *Whiteness: The communication of social identity* (pp. 264–78). Thousand Oaks, CA: Sage.

Stierlin, H. (1981). The parents' Nazi past and the dialogue between generations. *Family Process, 20,* 379–90.

Stierlin, H. (1993). The dialogue between the generations about the Nazi era. In B. Heimannsberg & C. J. Schmidt (Eds.), *Collective silence: German identity and the legacy of shame* (pp. 143–61). San Francisco: Jossey-Bass.

St. Louis *Post-Dispatch* (2004, September 24). A dangerous trend [editorial], p. B08.

Sullivan, S. (2003). Remembering the gift: W. E. B. DuBois on the unconscious and economic operations of racism. *Transactions of the Charles S. Peirce Society, 39,* 205–25.

Sullivan, S. (2006). *Revealing whiteness: The unconscious habits of racial privilege.* Bloomington: Indiana University Press.

Szasz, T. (1976). *Schizophrenia.* New York: Basic Books.

Tatum, B. D. (1997). *"Why are all the black kids sitting together in the cafeteria?" And other conversations about race.* New York: Basic Books.

Taylor, P. C. (2004). Silence and sympathy: Dewey's whiteness. In G. Yancy (Ed.), *What white looks like: African-American philosophers on the whiteness question* (pp. 227–41). New York: Routledge.

Thandeka (1999). *Learning to be white: Money, race, and God in America.* New York: Continuum.

Thorngate, W. (1988). On paying attention. In W. J. Baker, L. P. Mos, H. V. Rappard, & H. J. Stam (Eds.), *Recent trends in theoretical psychology* (pp. 247–63). New York: Springer Verlag.

Thussu, D. K. (2004). Media plenty and the poverty of news. In C. Paterson & A. Sreberny (Eds.), *International news in the 21st century* (pp. 47–61). Eastleigh, United Kingdom: John Libbey Publishing for University of Luton Press.

Tovares, A. V. (2007). Family members interacting while watching TV. In D. Tannen, S. Kendall, & C. Gordon (Eds.), *Family talk: Discourse and identity in four American families* (pp. 283–309). New York: Oxford University Press.

Townley, C. (2006). Toward a revaluation of ignorance. *Hypatia, 21*(3), 37–55.

Trepper, T. S., & Barrett, M. J. (1989). *Systemic treatment of incest: A therapeutic handbook*. New York: Brunner/Mazel.

Tuana, N. (2006). The speculum of ignorance: The women's health movement and epistemologies of ignorance. *Hypatia, 21*(3), 1–19.

Tully, S. R. (1995). A painful purgatory: Grief and the Nicaraguan mothers of the disappeared. *Social Science and Medicine, 40,* 1597–610.

Ulrich, W. (2002). Boundary critique. In H. G. Daellenbach & R. L. Flood (Eds.), *The informed student guide to management science* (pp. 41–42). London: Thomas Learning.

Ulrich, W. (2003). Beyond methodology choice: Critical systems thinking as critically systemic discourse. *Journal of the Operational Research Society, 54,* 325–42.

United States House of Representatives (1977). Hearings Before the Subcommittee on International Organizations of the Committee on International Relations (Rep. Donald M. Fraser, Chairman), Human Rights in East Timor and the Question of the Use of U.S. Equipment by the Indonesian Armed Forces, June 28 and July 19, 1977, 95th Congress, 1st Session, Washington, DC: U.S. Government Printing Office.

Ussher, J. M. (2004). Premenstrual syndrome and self-policing: Ruptures in self-silencing leading to increased self-surveillance and blaming of the body. *Social Theory and Health, 2,* 254–72.

Walker, A. J. (1996). Couples watching television: Gender, power, and the remote control. *Journal of Marriage and the Family, 58,* 813–23.

Wamboldt, F. W., & Wolin, S. J. (1988). Reality and myth in family life: Changes across generations. *Journal of Psychotherapy and the Family, 4*(3–4), 141–65.

Weinshall, M. (2003). Means, ends, and public ignorance in Habermas's theory of democracy. *Critical Review, 15,* 23–58.

Weissman, R. (2002). Shredded: Justice for BAT. *Multinational Monitor, 23* (5), 6–7.

Westerman, W. (1994). Central American refugee testimonies and performed life histories in the sanctuary movement. In R. Benmayor & A. Skotnes (Eds.), *International yearbook of oral history and life stories, vol. III. Migration and identity* (pp. 167–81). New York: Oxford University Press.

Weston, K. (1991). *Families we choose: Lesbians, gays, kinship*. New York: Columbia University Press.

Wetzel, N. A., & Winawer, H. (1986). The psychosocial consequences of the nuclear threat from a family systems perspective. *International Journal of Mental Health, 15*(1–3), 298–313.

Willits, W. L. (1986). Pluralistic ignorance in the perception of parent-youth conflict. *Youth and Society, 18,* 150–61.

Woodward, C. (2006, November 27). Truth is hard to find behind doublespeak. Seattle *Post-Intelligencer*. Retrieved March 16, 2007, from http://seattlepi.com/ national/293725_doublespeak27.html.

Wright, S. (1991). Family effects of offender removal from the home. In M. Q. Patton (Ed.), *Family sexual abuse: Frontline research and evaluation* (pp. 135–46). Newbury Park, CA: Sage.

Wright, S. E., & Rosenblatt, P. C. (1987). Isolation and farm loss: Why neighbors may not be supportive. *Family Relations, 36,* 391–95.

Yancy, G. (2004). Introduction: Fragments of a social ontology of whiteness. In G. Yancy (Ed.), *What white looks like: African-American philosophers on the whiteness question* (pp. 1–23). New York: Routledge.

Yang, J.-H. (2004). Constraints on environmental news production in the U.S.: Interviews with American journalists. *Journal of International and Area Studies, 11*(2), 89–105.

Yaniv, I., Benador, D., & Sagi, M. (2004). On not wanting to know and not wanting to inform others: Choices regarding predictive genetic testing. *Risk, Decision and Policy, 9,* 317–36.

Zerubavel, E. (2006). *The elephant in the room: Silence and denial in everyday life.* New York: Oxford University Press.

Zhang, A. Y., & Siminoff, L. A. (2003). Silence and cancer: Why do families and patients fail to communicate? *Health Communication, 15,* 415–29.

Zhao, Y. (2006). Are we fixing the wrong things? *Educational Leadership, 63*(8), 28–31.

Zimmerman, C. (2004). Denial of impending death: A discourse analysis of the palliative care literature. *Social Science and Medicine, 59,* 1769–80.

Zuk, G. H. (1965). On the pathology of silencing strategies. *Family Process, 4,* 32–49.

Index

179